生物工程与生物技术专业英语

English Course for Bioengineering and Biotechnology

主编　田英华　姜　彦
主审　刘晓兰

哈尔滨工程大学出版社

内容简介

本书选择了生物工程和生物技术的原理、发展和应用等方面的专业知识,以生物工程学、基因工程、酶学等内容为主,力求使读者能够接触更多的专业词汇、形式多样的文体和更多实用句型。主要内容:第1单元为氨基酸的生产方法、用途和前景,以及各种氨基酸的生产工艺;第2单元为酶学的发展史以及纤维素酶的应用;第3单元为奥地利生物工程的发展史和基因的简介及应用;第4单元为食品生物技术的代表性产品及食品和饲料添加剂的生产;第5单元为固态发酵的原理、特点及固态发酵的应用;第6单元为固定化酶和细胞的原理、方法及固定化酶和细胞的应用;第7单元为生物工程下游技术中的利用反向微团技术提取蛋白质,以及植物细胞培养中次级代谢产物的提取;第8单元为抗体生产的发展、方法、未来趋势及生化工程的研究进展。

本书可作为生物工程及生物技术专业英语教材,也可作为生物工程及生物技术相关人员学习英语的参考书。

图书在版编目(CIP)数据

生物工程与生物技术专业英语:英文/田英华,姜彦主编. —哈尔滨:哈尔滨工程大学出版社,2017.3(2023.7 重印)
ISBN 978 - 7 - 5661 - 1471 - 6

Ⅰ. ①生… Ⅱ. ①田… ②姜… Ⅲ. ①生物工程 - 英语 - 高等学校 - 教材 Ⅳ. ①Q

中国版本图书馆 CIP 数据核字(2017)第 048459 号

出版发行	哈尔滨工程大学出版社
社　　址	哈尔滨市南岗区南通大街 145 号
邮政编码	150001
发行电话	0451 - 82519328
传　　真	0451 - 82519699
经　　销	新华书店
印　　刷	哈尔滨午阳印刷有限公司
开　　本	787 mm×960 mm　1/16
印　　张	12.5
字　　数	345 千字
版　　次	2017 年 3 月第 1 版
印　　次	2023 年 7 月第 6 次印刷
定　　价	31.80 元

http://www.hrbeupress.com
E-mail:heupress@ hrbeu.edu.cn

PREFACE

English Course for Bioengineering and Biotechnology

  专业英语课程的目的是通过课程的学习与训练，使学生掌握常用专业词汇和科技英语表达方式，以提高对科技英语的阅读理解能力。本教材的编写原则是：有利于学生通过专业知识学习英语；教材精心选取了与生物工程和生物技术相关的科技成果信息和报道，涉及当前经典领域，覆盖面广，主要包括生物工程基础及概论、酶学、基因工程、生物技术、发酵工程及生物化学理论与实验技术等，代表性强，便于学生通过专业知识学习英语，了解当前生物工程发展的状况及趋势；课文难度略难于科普读物，便于学习和讲授。本教材提供了具有专业特色的英语表达形式和科技英语常用句式，有助于学生掌握阅读和翻译专业文献的技巧。

  本教材的特点是：均选自英文原版书籍；提高阅读理解能力；包含了科技英语中的主要语法、词与词组；重视词汇；重视写作能力的培养。

  本书的各个单元均设有 A,B 两部分，可以根据学生及课程的实际情况讲授全部或部分内容。

  本书可作为生物工程及生物技术专业英语教材，也可作为生物工程及生物技术相关人员学习英语的参考书。

  本书第 1 至 2 单元由姜彦编写；第 3 至 8 单元由田英华编写，全书由刘晓兰主审。

  由于编者水平有限，书中疏漏之处在所难免，敬请专家和广大读者提出宝贵意见。

编 者
2016 年 10 月

English Course for Bioengineering and Biotechnology CONTENTS

Unit 1 1
- Part A Amino Acids 1
- Part B Production of Amino Acid 21

Unit 2 38
- Part A Brief History of Enzymology 38
- Part B Applications of Cellulase Enzymes 54

Unit 3 64
- Part A History of Biotechnology in Austria 64
- Part B Genetic Engineering 83

Unit 4 95
- Part A Food Biotechnology 95
- Part B Production of Food and Feed Additives 104

Unit 5 109
- Part A Solid-state Fermentation 109
- Part B Applications of Solid-state Fermentation 117

Unit 6 122
- Part A Immobilization of Enzymes and Cells 122
- Part B Applications of Immobilized Enzymes and Cells 137

Unit 7 146
- Part A Protein Extraction Using Reverse Micelles 146
- Part B Gas Concentration Effects on Secondary Metabolite Production by Plant Cell Cultures 164

Unit 8 171
- Part A Established Bioprocesses for Producing Antibodies 171
- Part B The Progress in Biochemical Engineering 181

References 192

Unit 1

Part A

Amino Acids

1 Introduction

Proteins are **polymers** of amino acids, with each amino acid residue joined to its **neighbor** by a specific type of **covalent bond**. (The term "residue" reflects the loss of the elements of water when one amino acid is joined to another.) Proteins can be broken down (**hydrolyzed**) to their **constituent** amino acids by a variety of methods, and the earliest studies of proteins naturally focused on the free amino acids derived from them. Twenty different amino acids are commonly found in proteins. The first to be discovered was **asparagine**, in 1806. The last of the 20 to be found, **threonine**, was not identified until 1938. All the amino acids have trivial or common names, in some cases **derived from** the source from which they were first isolated. Asparagine was first found in **asparagus**, and **glutamate** in wheat gluten; **tyrosine** was first isolated from cheese (its name is derived from the *Greek tyros*, "cheese"); and **glycine** (*Greek glykos*, "sweet") was so named because of its sweet taste.

Amino Acids Share Common Structural Features

All 20 of the common amino acids are α-amino acids. They have a **carboxyl** group and an amino group bonded to the same carbon **atom** (the α carbon) (Fig. 1.1). They differ from each other in their side chains, or R groups, which vary in structure, size, and electric charge, and which influence the solubility of the amino acids in water. In addition to these 20 amino acids there are many less common ones. Some are residues modified after a protein has been **synthesized**; others are amino acids present in living organisms but not as constituents of proteins. The common amino acids of proteins have been assigned three-letter **abbreviations** and one-letter symbols (Tab. 1.1), which are used as shorthand to indicate the composition and sequence of amino acids polymerized in proteins.

Key Convention: The three-letter code is transparent, the abbreviations generally consisting of the first three letters of the amino acid name. The one-letter code was devised by *Margaret Oakley Dayhoff* (1925—1983), considered by many to be the founder of the field of *bioinformatics*. The one-letter code reflects an attempt to reduce the size of the data files (in an era of punch card computing) used to describe amino acid sequences. It was designed to be easily memorized, and understanding its origin can help students do just that.

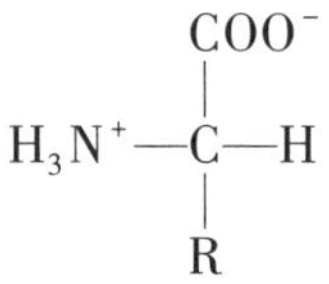

$$\begin{array}{c} COO^- \\ | \\ H_3N^+ - C - H \\ | \\ R \end{array}$$

Fig. 1.1 General structure of an amino acid. This structure is common to all but one of the α-amino acids. (Proline, a cyclic amino acid, is the exception.) The R group, or side chain, attached to the α carbon is different in each amino acid.

Tab. 1.1 Properties and Conventions Associated with the Common Amino Acids Found in Proteins

Amino acid	Abbreviation/ Amino acid symbol		Mr	pK_a values			pI	Hydropathy index	Occurrence in proteins(%)
				pK_1 (—COOH)	pK_2 (—NH_3^+)	pK_R (R group)			
Nonpolar, aliphatic R groups									
Glycine	Gly	G	75	2.34	9.60		5.97	−0.4	7.2
Alanine	Ala	A	89	2.34	9.69		6.01	1.8	7.8
Proline	Pro	P	115	1.99	10.96		6.48	1.6	5.2
Valine	Val	V	117	2.32	9.62		5.97	4.2	6.6
Leucine	Leu	L	131	2.36	9.60		5.98	3.8	9.1
Isoleucine	Ile	I	131	236	9.68		6.02	4.5	5.3
Methionine	Met	M	149	2.28	9.21		5.74	1.9	2.3

Tab. 1.1 (Continued)

Amino acid	Abbreviation/ Amino acid symbol		Mr	pK_a values			pI	Hydropathy index	Occurrence in proteins (%)
				pK_1 (—COOH)	pK_2 (—NH_3^+)	pK_R (R group)			
Aromatic R groups									
Phenylalanine	Phe	F	165	1.83	9.13		5.48	2.8	3.9
Tyrosine	Tyr	Y	181	2.20	9.11	10.07	5.66	−1.3	3.2
Tryptophan	Trp	W	204	2.38	9.39		5.89	−0.9	1.4
Polar, uncharged R groups									
Serine	Ser	S	105	2.21	9.15		5.68	−0.8	6.8
Threonine	Thr	T	119	2.11	9.62		5.87	−0.7	5.9
Cysteine	Cys	C	121	1.96	10.28	8.18	5.07	2.5	1.9
Asparagine	Asn	N	132	2.02	8.80		5.41	−3.5	4.3
Glutamine	Gln	Q	146	2.17	9.13		5.65	−3.5	4.2
Positively charged R groups									
Lysine	Lys	K	146	2.18	8.95	10.53	9.74	−3.9	5.9
Histidine	His	H	155	1.82	9.17	6.00	7.59	−3.2	2.3
Arginine	Arg	R	174	2.17	9.04	12.48	10.76	−4.5	5.1
Negatively charged R groups									
Aspartate	Asp	D	133	1.88	9.60	3.65	2.77	−3.5	5.3
Glutamate	Glu	E	147	2.19	9.67	4.25	3.22	−3.5	6.3

For all the common amino acids except glycine, the α-carbon is bonded to four different groups: a carboxyl group, an amino group, an R group, and a **hydrogen atom** (Fig. 1.1; in glycine, the R group is another hydrogen atom). The α-carbon atom is thus a **chiral center**. Because of the **tetrahedral** arrangement of the **bonding orbitals** around the α-carbon atom, the four different groups can occupy two unique **spatial** arrangements, and thus amino acids have two possible **stereoisomers**. Since they are **non-superposable** mirror images of each other (Fig. 1.2), the two forms represent a class of stereoisomers called **enantiomers**. All molecules with a chiral center are also **optically active**—that is, they rotate **plane-polarized light**.

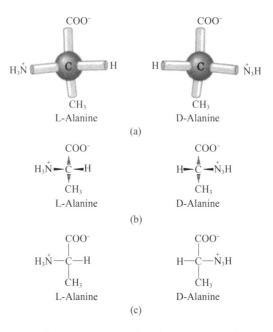

Fig. 1.2 Stereoisomerism in α-amino acids.
(a) The two stereoisomers of alanine, L-and D-alanine are non-superposable mirror images of each other (enantiomers). (b) (c) Two different conventions for showing the configurations in space of stereoisomers. In perspective formulas (b) the solid wedge-shaped bonds project out of the plane of the paper, the dashed bonds behind it. In projection formulas (c) the horizontal bonds are assumed to project out of the plane of the paper, the vertical bonds behind. However projection formulas are often used casually and are not always intended to portray a specific stereochemical configuration.

Unit 1

New Words

amino acid	氨基酸
protein	蛋白质
polymer	聚合物,聚合体,高聚物
neighbor	邻居
covalent bond	共价键
hydrolyze	水解
constituent	构成的,组成的
asparagines	天冬酰胺,天门冬素,氨羰丙氨酸
threonine	苏氨酸,羟丁胺酸
derive from	源出,来自,得自,衍生于
asparagus	芦笋,龙须菜,天冬
glutamate	谷氨酸,谷氨酸盐
tyrosine	酪氨酸
glycine	甘氨酸,氨基乙酸
carboxyl	羧基
atom	原子
synthesize	合成,综合
abbreviation	缩写,缩写词,简称
devise	设计,想出,发明,图谋
bioinformatics	生物信息学,生物信息,生物资讯
hydrogen atom	氢原子
chiral center	手性中心
tetrahedral	四面体的,有四面的
bonding orbital	成键轨函,成键轨道,成键轨函数
spatial	空间的,存在于空间的,受空间条件限制的
stereoisomer	立体异构体
non-superposable	不可叠加的,不可重合的
enantiomer	对映体,对映异构体
optically active	光学活性的,有旋光力的,起偏振作用的
plane-polarized	平面偏振光,平面偏光

2 The Amino Acid Residues in Proteins Are L Stereolsomers

Nearly all biological compounds with a chiral center occur naturally in only one stereoisomeric form, either D or L. The **amino acid residues** in protein **molecules** are **exclusively** L stereoisomers. D-Amino acid residues have been found in only a few, generally small **peptides**, including some peptides of bacterial cell walls and certain peptide **antibiotics**.

It is remarkable that virtually all amino acid residues in proteins are L stereoisomers. When chiral compounds are formed by **ordinary** chemical reactions, the result is a **racemic** mixture of D and L isomers, which are difficult for a chemist to distinguish and separate. But to a living system, D and L isomers are as different as the right hand and the left. The formation of stable, repeating substructures in proteins generally requires that their constituent amino acids be of one **stereochemical** series. Cells are able to specifically synthesize the L isomers of amino acids because the active sites of enzymes are **asymmetric**, causing the reactions they catalyze to be **stereospecific**.

New Words

amino acid residue	氨基酸残基
molecule	分子
exclusively	唯一地,专有地,排外地
peptide	肽
antibiotics	抗生素,抗菌药物
ordinary	普通的,平凡的,平常的
racemic	外消旋的
stereochemical	立体化学的
asymmetric	不对称的,非对称的
stereospecific	立体定向的,立体专一性的

Unit 1

3 Amino Acids Can Be Classified by R Group

Knowledge of the chemical properties of the common amino acids is central to an understanding of **biochemistry**. The topic can be simplified by grouping the amino acids into five main classes based on the properties of their R groups, in particular, their **polarity**, or tendency to **interact with** water at biological pH (near pH 7.0). The polarity of the R groups varies widely, from **nonpolar** and **hydrophobic** (water-insoluble) to highly **polar** and hydrophilic (water-soluble).

The structures of the 20 common amino acids are shown in Fig. 1.3, and some of their properties are listed in Tab. 1.1. Within each class there are gradations of polarity, size, and shape of the R groups.

Nonpolar, Aliphatic R Groups

The R groups in this class of amino acids are nonpolar and hydrophobic. The side chains of alanine, valine, leucine, and isoleucine tend to cluster together within proteins, stabilizing protein structure by means of hydrophobic interactions. Glycine has the simplest structure. Although it is most easily grouped with the nonpolar amino acids, its very small side chain makes no real contribution to hydrophobic interactions. Methionine, one of the two sulfur-containing amino acids, has a nonpolar thioether group in its side chain. Proline has an aliphatic side chain with a distinctive cyclic structure. The secondary amino(imino) group of proline residues is held in a rigid conformation that reduces the structural flexibility of polypeptide regions containing proline.

Aromatic R Groups Phenylalanine, Tyrosine, and Tryptophan, with their aromatic side chains, are relatively nonpolar(hydrophobic). All can participate in hydrophobic interactions. The hydroxyl group of tyrosine can form hydrogen bonds, and it is an important functional group in some enzymes. Tyrosine and tryptophan are significantly more polar than phenylalanine, because of the tyrosine hydroxyl group and the nitrogen of the tryptophan indole ring.

Tryptophan and tyrosine, and to a much lesser extent phenylalanine, absorb ultraviolet light (Fig. 1.4). This accounts for the characteristic strong absorbance of light by most proteins at a wave length of 280 nm, a property exploited by researchers in the characterization of proteins.

Polar, Uncharged R Groups

The R groups of these amino acids are more soluble in water, or more hydrophilic, than those of the nonpolar amino acids, because they contain functional groups that form hydrogen bonds with water. This class of amino acids includes serine, threonine, cysteine, asparagine, and glutamine.

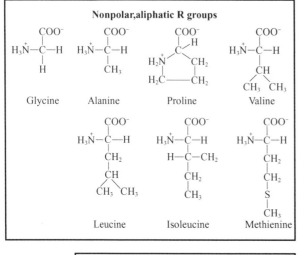
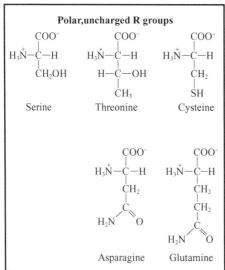
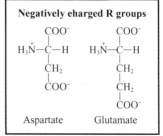

Fig. 1.3   The 20 common amino acids of proteins

The structural formulas show the state of ionization that would predominate at pH 7.0. The unshaded portions are those common to all the amino acids; the portions shaded in pink are the R groups

The polarity of serine and threonine are contributed by their hydroxyl groups; that of cysteine by its sulfhydryl group, which is a weak acid, and can make weak hydrogen bonds with oxygen or nitrogen; and that of asparagines and glutamine by their amide groups.

Unit 1

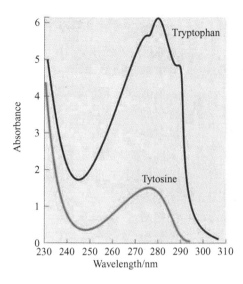

Fig. 1.4 Absorption of ultraviolet light by aromatic amino acid

Positively Charged (Basic) R Groups

The most hydrophilic R groups are those that are either positively or negatively charged. The amino acids in which the R groups have significant positive charge at pH 7.0 are lysine, which has a second primary amino group at the ε position on its aliphatic chain; arginine, which has a positively charged guanidinium group; and histidine, which has an aromatic imidazole group. As the only common amino acid having an ionizable side chain with pK_a near neutrality, histidine may be positively charged (protonated form) or uncharged at pH 7.0. Its residues facilitate many enzymes-catalyzed reactions by serving as proton donors/acceptors.

Negatively Charged (Acidic) R Groups

The two amino acids having R groups with a net negative charge at pH 7.0 are aspartate and glutamate, each of which has a second carboxyl group.

New Words

polarity	极性,两极,对立
hydrophobic	疏水的,狂犬病的,恐水病的
polar	两极的

alanine	丙氨酸
valine	缬氨酸
leucine	亮氨酸
glycine	甘氨酸,氨基乙酸
methionine	蛋氨酸,甲硫氨酸
sulfur-containing	含硫的
proline	脯氨酸
polypeptide	多肽,缩多氨酸
aromatic	芳香的,芬芳的,芳香族的
phenylalanine	苯基丙氨酸
tyrosine	酪氨酸
tryptophan	色氨酸
hydrophobic interaction	疏水作用
hydroxyl group	羟基
hydrogen bond	氢键
indole ring	吲哚环
ultraviolet light	紫外线,紫外辐射,紫外光
sulfhydryl group	巯基
amide group	酰胺基
guanidinium group	胍基
histidine	组氨酸
imidazole group	咪唑,异吡唑
ionizable	可电离的
protonated	质子化的

4 Glutamaic acid

The story of amino acid production started in 1908 when the chemist, Dr K. Ikeda, was working on the **flavouring** components of **kelp**. Kelp is traditionally very popular with the Japanese due to the specific taste of its preparations, kombu and katsuobushi (Fig. 1.5). After acid **hydrolysis** and **fractionation** of kelp, Dr K. Ikeda discovered that one specific fraction he had isolated consisted of **glutamic acid**, which after **neutralization** with **caustic soda**, developed an entirely new, delicious taste. This was the birth of the use of **monosodium glutamate** (**MSG**) as a flavour-enhancing compound, the production of monosodium glutamate was soon commercialized by the Ajinomoto company based on its isolation from vegetable proteins such as **soy** or wheat protein. Since less than 1 kg MSG could be isolated from 10 kg of raw material. The waste fraction was high. The chemical synthesis of D, L-Glutamate, which had been partially successful, was also of little use since the **sodium salt** of the D-**Isomer** is tasteless.

The **breakthrough** in the production of MSG was the isolation of a specific **bacterium** by Dr S. Udaka and Dr S. Kinoshita at Kyowa Hakko kogyo in 1957. They screened for **amino acid**-excreting **microorganisms** and discovered that their isolate, No. 534, had grown on a **mineral salt** medium excreted L-Glutamate. It soon became apparent that the isolated organism needed **biotin** and that L-Glutamate Excretion was triggered by an insufficient supply of biotin. A number of **bacteria** with similar properties were also isolated, which are today all known by the species name **corynebacterium glutamicum** (c. glutamicum for short) (Fig. 1.6). c. glutamicum is a **gram-positive** bacterium, which can be isolated from soil. Together with genera like **Streptomycetes**, **propionibacterium** or **Arthrobacter**, it belongs to the **actinomycetes** subdivision of gram-positive bacteria. The successful commercialization of MSG production with this bacterium provided a big boost for amino acid production with c. glutamicum and later with other bacteria like **e. coli** as well. **Nucleotide** production for use as **flavour enhancers** also developed rapidly in the 1970s with **c. ammonia genes**, which is closely related to **c. glutamicum**. The production **mutants** and the processed developed also resulted in a demand for sophisticated fermentation devices. Consequently, the development of amino acid technology was an incentive for the fermentation industry in general.

Fig. 1.5 The ideogram for kombu as it appears on kelp preparation used as a food component

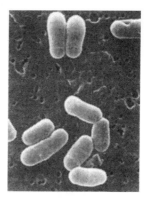

Fig. 1.6 Electron micrograph of corynebacterium glutamicum showing the typical V-shape of two cells as a consequence of cell division

New Words

flavouring	调味料
kelp	海藻,海藻灰(可提取碘的)
hydrolysis	水解
fractionation	分馏法
glutamic acid	谷氨酸
neutralization	中和
caustic soda	苛性钠
monosodium glutamate (MSG)	味精,味素;谷氨酸一钠(味精的化学成分)
soy	酱油,大豆
sodium salt	钠盐;(专指)氯化钠
isomer	异构体
breakthrough	突破
bacterium (bacteria)	细菌
amino acid	氨基酸
microorganism	微生物,微小动植物
mineral salt	天然盐
biotin	维生素H,生物素
bacteria	细菌
corynebacterium	棒状杆菌
corynebacterium glutamicum	谷氨酸棒杆菌
gram-positive	革兰氏(染色)阳性
genera	类,属
streptomycete	链霉菌
propionibacterium	丙酸杆菌
arthrobacter	节杆菌属
actinomycete	放射菌类
e. coli	大肠杆菌
nucleotide	核苷
flavour enhancer	香味增强剂,风味增强剂,鲜味增强剂(如味精等)
mutant	突变体

Unit 1

Notes

After acid hydrolysis and fractionation of kelp, Dr K. Ikeda discovered that one specific fraction he had isolated consisted of glutamic acid, which after neutralization with caustic soda, developed an entirely new, delicious taste. 海藻经酸水解和分馏后，Ikeda 博士发现他所分离的包含谷氨酸的特定馏分经过苛性钠中和后，逐渐形成了全新的可口的味道。

❺ Commerical Use of Amino Acids

Amino acids are used for a variety of purposes. The food industry requires L-Glutamate as a flavour enhancer, and **glycine** as a **sweetener** in juice, for instance (Tab. 1.2). The chemical industry requires amino acids as building blocks for a diversity of compounds. The **pharmaceutical** industry requires the amino acids themselves in **infusions** in particular the essential amino acids or in special dietary food. And last but not least, a large market for amino acids is their use as animal feed **additive**. The reason is that typical **feedstuffs**, such as **soybean meal** for pigs, are poor in some essential amino acids, like **methionine**, for instance. This is illustrated in Fig. 1.7 where the nutritive value of soybean meal is given by the barrel but the use of the total barrel is limited by the stave representing methionine. Methionine is added for this reason, and considerably increases the **effectiveness** of the feed. The addition of as little as 10 kg methionine per tonne increase the protein quality of the feed just as effectively as adding 160 kg soybean meal or 56 kg fish meal. The first limiting amino acid in feed based on crops and **oil seed** is usually L-Methionine, followed by L-**Lysine**, and L-**Threonine**. Another aspect of feed **supplementation** is that with a balanced amino acid content the manure contains less **nitrogen** thus reducing environmental pollution.

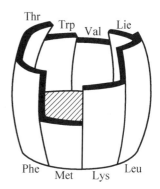

Fig. 1.7 The barrel represents the nutritive value of soybean meal, which is first limited by its methionine content

Tab. 1.2 Current amounts of amino acids produced

Production scale (tonnes y^{-1})	Amino acid	Preferred production method	Main use
800,000	L-Glutamic acid	Fermentation	Flavour enhancer
350,000	L-Lysine	Fermentation	Feed additive
350,000	D,L-Methionine	Chemical synthesis	Feed additive
10,000	L-Aspartate	Enzymatic catalysis	Aspartame
10,000	L-Phenylalanine	Fermentation	Aspartame
15,000	L-Threonine	Fermentation	Feed additive
10,000	Glycine	Chemical synthesis	Feed additive, sweetener
3,000	L-Cysteine	Reduction of cystine	Feed additive, pharmaceutical
1,000	L-Arginine	Fermentation, extraction	Pharmaceutical
500	L-Leucine	Fermentation, extraction	Pharmaceutical
500	L-Valine	Fermentation, extraction	Pesticides, pharmaceutical
300	L-Tryptophan	Whole cell process	Pharmaceutical
300	L-Isoleucine	Fermentation, extraction	Pharmaceutical

Over the years the demand for amino acids has increase dramatically. The market is growing steadily by about 5 to 10 per cent per year. Thus, within 10 years the total market has approximately doubled (Fig. 1.8). Some amino acids, such as L-Lysine, which is required as a feed additive, display a particularly great increase. The world market for this amino acid has increased more than 20-fold in the past two decades. Other amino acids have appeared on the market, like L-Threonine, L-**Aspartate** or L-**Phenylalanine**, the latter two being required for the synthesis of the newly developed sweetener **aspartame**. Estimates for current worldwide demand for the most relevant amino acids are given in Table 1.1. L-**Glutamate** continues to occupy the top position followed by L-Lysine together with D,L-Methionine, while the other amino acids trail behind at a considerable distance.

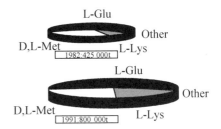

Fig. 1.8 The amino acid market doubles about every ten years (t = tonnes)

There is a close interaction between the prices of the amino acids and the **dynamics** of the market. More efficient fermentation technology can provide cheaper products and hence boost demand. This in turn will lead to production on a larger scale with a further **reduction of costs**. However, since the supply of some amino acids, e. g. L-Lysine, as a feed additive is directly competitive with soybean meal (the **natural** L-Lysine **source**) there are considerable fluctuations in the amino acid demand depending on the crop yields. The amino acids produced in the largest quantities are also the cheapest (Fig. 1.9). The low prices in turn dictate the location of the production plants. The main factors governing the location of production plants are the price of the **carbon source** and the local market. Large L-Glutamate production plants are spreading all over the world, with a significant presence in the Far East, e. g. Thailand and Indonesia. For L-Lysine the situation is different. Since one-third of the world market is in North America and there is convenient access to **maize** as a feedstock material for the fermentation process, about one-third of the L-Lysine production capacity is located there. In almost all cases, the companies producing L-Lysine are associated with the maize **milling** industry, either as producers, in joint ventures or as suppliers of cheap sugar. This illustrates the fact that the commercial production of amino acids is a vigorously growing and changing field with many global interactions.

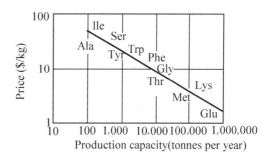

Fig. 1.9 The amino acids with the largest market are the cheapest

New Words

sweetener	(调味用)甜料,甜味佐料
pharmaceutical	药物;制药(学)上的
infusion	溶液;注射
additive	添加剂
feedstuff	饲料,饲料中的营养成分
soybean	大豆
methionine	蛋氨酸,甲硫氨酸
meal	粗磨粉,颗粒物
effectiveness	效力

oil seed	含油种子
lysine	赖氨酸
supplementation	增补,补充,追加
nitrogen	氮
aspartate	天(门)冬氨酸盐(或酯)
phenylalanine	苯基丙氨酸
aspartame	天(门)冬氨酰苯丙氨酸甲酯(一种约比蔗糖甜200倍的甜味剂)
dynamic	动态
reduction of cost	降低成本
natural source	天然源
carbon source	碳源
maize	玉米
milling	磨,制粉

6 Production Methods and Tools

Some amino acids are chemically synthesized, such as glycine, which has no **stereochemical** centre, or D,L-Methionine. This latter **sulphur**-containing amino acid can be added to feed as a **racemic mixture**, since animals contain a D-Amino acid **oxidase**, which together with a **transaminase** activity, converts D-Methionine to the nutritively effective L-Form. The classical procedure of amino acid isolation from acid **hydrolysates** of proteins is still in use for selected amino acids with a low market volume, e.g. L-**Cysteine** (Tab. 1.2). Other methods in use are those of precursor conversion with bacteria, or enzymatic synthesis. However, for L-Amino acids required in large volumes, fermentation production with bacteria is the method of choice.

1 Classical strain development

However, bacteria do not normally excrete amino acids in significant amounts because **regulatory mechanisms** control the amino acid synthesis in an economical way. Therefore, mutants have to be generated which over-synthesis the respective amino acid. A large number of amino acid—producing bacteria have been derived by **mutagenesis** and screening programmes. This has involved the consecutive application of:

Unit 1

undirected mutagenesis;

selection for a specific phenotype;

selection of the mutant with the best amino acid accumulation.

Taking the best resulting strain, the entire procedure was repeated over several additional rounds to increase the productivity each time, and, eventually, resulted in an industrial producer (see Tab. 1.3 as an example). Due to this **optimization** over several decades, together with the accompanying process adaptation excellent high-performance strains are now available. The certainly carry a variety of unknown mutations also decisive for their production properties, as will become evident from the examples described below.

Tab. 1.3 A genealogy of strains obtained by classical mutagenesis and screening, showing the yield improvement obtained and some phenotypic characters known

Strain	Character	Yield of L-Lysine (%)
AJ 1511	Wild type	0
AJ 3445	AECr	16
AJ 3424	AECr Ala-	33
AJ 3796	AECr Ala-CCLr	39
AJ 3990	AECr Ala-CCLr MLr	43
AJ 1204	AECr Ala- CCLr MLr FPs	50

2 Application of recombinant techniques

In conjunction with this classical technique for strain development, **recombinant** DNA techniques are also applied. They serve

to rapidly develop new producers by increasing limiting enzyme activities;

to analysis mechanisms of flux control;

to combine this knowledge with classically obtained stains for their further development.

3 Intracellular flux analysis

An exciting new approach in strain development combining both the genetic and classical procedure is the reliable **quantification** of the carbon **fluxes** in the living cell. A great deal of progress has been made here recently in developing to a high level of sophistication the old **isotope labeling technique. In particular, with** ^{13}C-**NMR spectroscopy** the **intracellular** fluxes were quantified to extreme high resolution. For instance, in c. glutamicum it has even been possible to quantify the exchange flux rates as are present in the **pentose phosphate** pathway. Such flux identifications are of major assistance in selecting the reactions in the central **metabolism** to be modified by genetic engineering.

4 Functional genomics

Another tool whose potential is only now being exploited is the **genome** analysis of producer strain. The availability of the entire sequence of the **chromosomes** from c. glutamicum and e. Coli opens up exciting possibilities to compare mutants and to uncover new mutations essential for high **overproduction** of metabolites. For instance, RNA analysis using chip technology will make it possible to detect whether a specific gene is altered in its expression for producers of different efficiency. New mutations and genes might thus be discovered which are not directly concerned with carbon fluxes, but rather with total cell control, or are involved in **energy metabolism**. Chip technology will also make it possible to use genome analysis as a tool to qualify individual fermentations, thus resulting in still further improvements and consolidations of the production processes.

New Words

stereochemical	立体化学的
sulphur	硫,硫黄
racemic mixture	外消旋混合物
oxidase	氧化酶
transaminase	(= aminotransferase)转氨酶
hydrolysate	水解产物
cysteine	半胱氨酸,巯基丙氨酸
enzymatic synthesis	酶催化合成
strain	品系
regulatory mechanism	调节机制
mutagenesis	突变形成,变异发生

Unit 1

phenotype	显型,表现型
optimization	最佳化,最优化
recombinant	重组,重组体;重组的
quantification	量(化)
flux	流量,通量
isotope	同位素
NMR	(= nuclear magnetic resonance)核磁共振
spectroscopy	光谱学,波谱学,分光镜使用
intracellular	细胞内的
pentose	戊糖
phosphate	磷酸盐
metabolism	新陈代谢
genomics	基因组学
genome	基因组,染色体组
chromosome	染色体
overproduction	生产过剩
energy metabolism	能量代谢

7 Outlook

Although amino acids are now among the classical products in **biotechnology**, their constant development means that processes must be improved, new processed established and our understanding of the exceptional capabilities of producer strains deepened. Just one example of **molecular** research is the recent discovery of the L-Lysine export carrier, which opens up an entirely new field in the metabolism of amino acids in bacteria in general. Moreover, much information has been gathered from strain development in conjunction with fermentation technology, with the new science of **metabolic engineering** at the interface between them. In fact, amino acid production is an outstanding example of the integration of many different techniques. In this way, the early Japanese activities on the taste of kelp laid the foundation for the continuing very successful and flourishing production of amino acids.

New Words

biotechnology	生物工艺学
molecular	分子的,由分子组成的
metabolic engineering	代谢工程

Part B

Production of Amino Acid

1 L-Glutamate

As already mentioned, L-Glutamate was the first amino acid to be produced. The very successful production still exclusively uses the original bacterium c. glutamicum. As metabolic pathways c. glutamicum uses glycolysis, the pentose phosphate pathway and the citric acid cycle to generate precursor metabolites and reduced pyridine nucleotides. However, this bacterium displays a special feature in the anaplerotic reactions of the citric acid cycle (Fig. 1.10). Since L-Glutamate is directly derived from α-ketoglutarate, a high capability for replenishing the citric acid cycle is, of course, a prerequisite for high glutamate production. It was originally assumed that only the phosphoenol pyruvate carboxylase is present as a carboxylating enzyme within the anaplerotic reactions. However, molecular research in close conjunction with ^{13}C-labelling studies and flux analysis showed that an additional carboxylating reaction must be present. The pursuit of this enzyme activity resulted in the detection of pyruvate carboxylase activity, (PyrC), and the cloning of its gene. This carboxylase was not detected by the original enzyme measurements since it is very unstable in crude extracts. Its detection requires an in situ enzyme assay using carefully permeabilised cells. Therefore, c. glutamicum has the pyruvate dehydrogenase (PyrDH) shuffling acetyl-CoA into the citric acid cycle but two enzymes supplying oxaloacetate: pyruvate carboxylase (PyrC) together with a phosphoenol pyruvate carboxylase (PEPC) (Fig. 1.10). The successful cloning of both genes together with mutant studies showed that both carboxylases can basically replace each other to ensure conversion of glucose-derived C3-units to oxaloacetate. This is different from e. coli, which has exclusively the phosphoenol pyruvate carboxylase serving this purpose, or Bacillus subtilis, where only the pyruvate carboxylase is present. Since c. glutamicum possesses both enzymes, it has an enormous flexibility for replenishing citric acid cycle intermediates upon their withdrawal.

The reductive amination of α-ketoglutarate to yield L-Glutamate is catalysed by glutamate dehydrogenase. The enzyme is a multimer, each subunite having a molecular weight of 49,100. It has a high specific activity of 1.8 mmol min^{-1} · mg protein, and L-Glutamate is present in the cell in a rather high concentration of about 150 mmol L^{-1}. In case of other amino acids, in contrast, the intracellular concentra-tions are usually below 10 mmol L^{-1}. The high concentration serves to ensure the supply of L-Glutamate directly required for cell synthesis and also for the supply of amino groups via transaminase reactions for a variety of cellular reactions. As much as 70% of the amino groups in cell material stems from L-Glutamate.

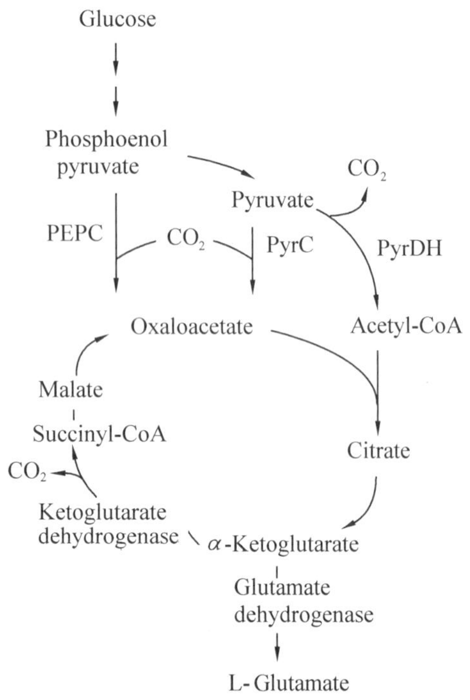

1 Production strains

For the biotechnological production of L-Glutamate the intracellularly synthesized

Fig. 1.10 Sketch of main reaction in c. glutamicum connected with the citric acid cycle and of relevance for L-Glutamate production. Abbreviations: PyrDH, pyruvate dehydrogenase; PyrC, pyruvate carboxylase; PEPC, phosphoenol pyruvate carboxylase

amino acid must be released from the cell. This is, of course, usually not the case since the charged L-Glutamate is retained by the cytoplasmic membrane, otherwise the cell would not be viable. However, as shown by the special circumstances in discovering c. glutamicum, L-Glutamate is already excreted when biotin is limiting. This striking fact is based on two essential characteristics:

A carrier is present mediating the active excretion of L-Glutamate;
The lipid environment of this carrier triggers its activity.

A specific carrier is required since otherwise, in addition to the charged L-Glutamate, other metabolites and ions would also leak from the cell, Moreover, only an active export enables the energy-dependent "uphill" transport of L-Glutamate from inside the cell (0.15 M) towards the very high concentrations obtained in fermentation broths (more than 1 M). However, for practical purposes, the triggering of active export by the appropriate molecular environment of the cytoplasmic membrane is important. The switches for tuning this environment and thus eliciting glutamate export are surprisingly diverse: (i) growth under biotin limitation, (ii) addition of local anaesthetics, (iii) addition of penicillin, (iv) addition of surfactants, (v) use of oleic acid auxotrophs, and (vi) use of glycerol auxotrophs. All of these means trigger L-Glutamate excretion. Although, overall, there are as yet no completely conclusive ideas on the molecular changes thus caused, nevertheless in the classical biotin effect part of the causal link to glutamate excretion is well understood. Biotin is a cofactor of the acetyl-CoA carboxylase. With limited supply, the activity of this enzyme is thus decreased and consequently the fatty acid synthesis is diminished. This leads to a decreased availability of phospholipids and a greatly decreased lipid to protein ratio in the membrane as well as a change in the degree of saturation of the fatty acids. Under biotin limitation the phospholipid content is drastically decreased from 32 to 17 mmol mg^{-1} dry weight, and the content of the unsaturated oleic acid increased relative to the saturated palmitic acid by 45%. This represents a severe alteration of the physical state of the membrane, which thus dramatically alters L-Glutamate efflux. The membrane composition is also affected in oleic acid and glycerol auxotrophic mutants. The use of such mutants enables the production of monosodium glutamate from substrates, which may be rich in biotin as well.

Apart from the export process and high glutamate dehydrogenase activity, another key reaction is that of α-ketoglutarate dehydrogenase (Fig. 1.11). This enzyme has a weak activity in c. glutamicum and it is also unstable. Therefore, under those conditions that result in glutamate efflux, the activity of this enzyme is also diminished. Exposing the cell to either penicillin, surfactants or biotin-limitation reduces the α-ketoglutarate dehydrogenase activity up to a residual activity of only 10%, whereas the activity of the glutamate dehydrogenase is hardly affected. The competing α-ketoglutarate dehydrogenase activity is therefore lowered, thus preventing an excess conversion of α-ketoglutarate to succinyl-CoA, and therefore favouring its conversion to L-Glutamate.

2 Production process

The most relevant factors influencing L-Glutamate formation are the ammonium concentration, the dissolved O_2 concentration and the pH. Although, in total, a large amount of ammonium is necessary for sugar conversion to L-Glutamate, a high concentration is inhibitory to growth as well to the production of L-Glutamate. Therefore, ammonium is added in a low concentration at the beginning of the fermentation and is then added continuously during the course of the fermentation. The oxygen concentration is controlled, since under conditions of insufficient oxygen, the production of L-Glutamate is poor and lactic acid as well as succinic acid accumulates, whereas with an excess oxygen supply the amount of α-ketoglutarate as a by-product accumulates. A low diagram of the process is shown in Fig. 1.11.

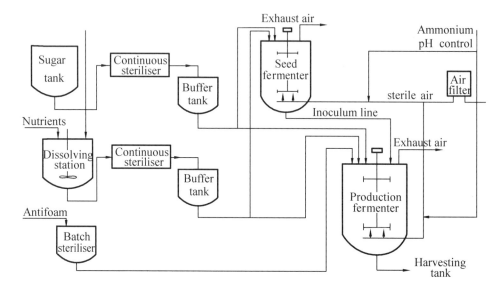

Fig. 1.11 A scheme of the material flow in an L-Glutamate production plant

For the actual fermentation the production strains are grown in fermenters as large as 500 m^3 (Fig. 1.12). After precultivation, the onset of L-Glutamate excretion is controlled by the addition of surfactants like polyoxyethylene sorbitan monopalmitate (Tween 40). Yields of 60% – 70% L-Glutamate, based on the glucose used, have been reported. At the end of the fermentation the broth contains L-Glutamate in the form of its ammonium salt. In a typical downstream process, the cells are separated and the broth is passed through a basic anion exchange resin. L-Glutamate anions will be bound to the resin and ammonia will be released. This ammonia can be recovered via distillation and reused in the fermentation. Elution is performed with NaOH to directly form MSG in the solution and to regenerate the basic anion exchanger. From the eluates, MSG may be crystallised directly followed by further conditioning steps like decolorisation and sieving to yield a food-grade quality.

Unit 1

Fig. 1.12 Amino acid production plant of Kyowa Hakko in Japan showing on the right seven large fermenters each 240 m^3 in size, suitable for L-Glutamate production

❷ L-Lysine

The second amino acid made exclusively with c. glutamicum, or its subspecies lactofermentum and flavum, is L-Lysine. The carbons of L-Lysine are derived in the central metabolism from pyruvate and oxaloacetate. In contrast to the special situation with L-Glutamate, where practically only a single reaction represents the synthesis pathway, L-Lysine is synthesized via a long pathway. Moreover, the first two steps of L-Lysine synthesis are shared with that of the other members of the aspartate family of amino acids: L-Methionine, L-Threonine and L-Isoleucine.

❶ The kinase initiating lysine synthesis is feedback-inhibited by lysine plus threonine

The first reaction initiating L-Lysine synthesis is catalysed by aspartate kinase. As is typical of an enzyme at the start of a lengthy synthesis pathway, aspartate kinase is controlled in its catalytic activity. The enzyme is inactive when L-Lysine plus L-Threonine together are present in excess, thus providing a feedback signal concerning the availability of these two major metabolites of the aspartate family of amino acids. The kinase has an interesting structure. It consists of 2 α-subunits of 421 amino acid residues each, and 2 β-subunits of 171 amino acid residues. An exciting discovery was that the amino acid sequence of the β-subunit is identical to that in the carboxy terminal part of the α-subunit. The molecular basis is that the gene for the smaller β-subunit, lysCβ, is an in-frame constituent part of the larger α-subunit. Thus two promoters are present at this locus: one driving lysCα expression together with that of the downstream gene, αsd, and one driving lysCβ and αsd expression. The regulatory features of the kinase reside in the β-subunit. Thus specifically altering the β-subunit structure, or those of both subunits together in their carboxy terminal part, results in a kinase which is no longer feedback regulated by L-Lysine plus L-Threonine. With such an insensitive kinase, c. glutamicum already excretes some L-Lysine, showing the rather simple type of flux control in this organism.

2 The synthase limits flux

A further important step of flux control within lysine biosynthesis is at the level of aspartate semialdehyde distribution. The dihydrodipicolinate synthase activity competes with the homoserine dehydrogenase for the aspartate semialdehyde. In c. glutamicum, the synthase is not regulated in its catalytic activity as is the corresponding enzyme in e. coli, for example. Instead, in c. glutamicum it is the amount of the protein which directly controls the flux. This is thus different from the kinase where the catalytic activity is regulated by L-Lysine and thereby controls the flux at a constant amount of protein. Graded overexpression of the synthase gene, dapA, has shown that with an increasing amount of synthase a graded flux increase towards L-Lysine is the result. Surprisingly, dapA overexpression also has a second consequence: the flux of aspartate semialdehyde into the branch leading to homoserine is already diminished with just two dapA copies. Due to the shortage of the homoserine-derived amino acids, this results in a weak growth limitation, which is advantageous for L-Lysine formation, since now more intermediates of the central metabolism are used for lysine synthesis instead for cell proliferation.

3 Lysine synthesis is split which ensures proper cell wall formation

A remarkable feature of c. glutamicum is its split pathway of L-Lysine synthesis. At the level of piperideine-2,6-dicarboxylate, flux is possible either via the 4-step succinylase variant or the 1-step dehydrogenase variant. In contrast, e. coli, for example, has only the succinylase variant and

Bacillus macerans only the dehydrogenase variant. The flux distribution via both pathways has been quantified in a study using NMR spectroscopy and [1-^{13}C] glucose as the substrate. Surprisingly the flux distribution is variable. Whereas at the start of the cultivation about three-quarters of the L-Lysine is made via the dehydrogenase variant, at the end the newly synthesized L-Lysine is almost exclusively made via the succinylase route. There is a mechanistic reason for this. As kinetic characterisations have shown, the dehydrogenase has a weak affinity towards its substrate, ammonium, with a K_m of 28 mmol L^{-1}. Thus at low ammonium concentrations, as are present at the end of the fermentation, the dehydrogenase cannot contribute to L-Lysine formation. Instead, flux via the succinylase variant is favoured, where after succinylation of piperideine-2,6-dicarboxylate, a transaminase incorporates the second amino group into the final L-Lysine molecule.

The key to understanding this luxurious pathway construction is provided by the amino acid D,L-Diaminopimelate. This amino acid is required for the synthesis of the activated muramyl peptide L-Ala-γ. D-Glu-D,L-Dap, which is one of the linking units in the peptidoglycan of the cell wall. Upon inactivation of the succinylase variant, a radical change to the cell morphology becomes apparent with low nitrogen supply. The cells are elongated, and furthermore less resistant to mechanical stress. If either the succinylase or the dehydrogenase variant is inactivated, L-Lysine accumulation is reduced to 40%. Thus both variants together ensure the proper supply of the crucial linking unit D,L-Diaminopimelate, as well as a high throughput for L-Lysine formation. The split pathway in c. glutamicum is an example of an important principle in microbial physiology: pathway variants are generally not redundant but evolved to provide key metabolites under different environmental conditions.

4 Export of L-Lysine

Amino acid transport has long been investigated in bacteria but, principally, this is only their import. In contrast, the molecular basis for amino acid export was completely unknown until 1996 since a specific export process appeared nonsensical. The breakthrough was achieved by the cloning of the lysine export carrier from c. glutamicum, which at one blow enabled amazing discoveries concerning the nature and relevance of a new type of exporter. The L-Lysine carrier, LysE, is a comparatively small membrane protein of 25.4 Da. It has the transmembrane spanning helices typical of carriers, but only five of them. A sixth hydrophobic segment is located between helix one and three and may dip into the membrane or be surface localised. Several distinct steps are involved in the translocation mechanism, which probably requires the dimerisation of LysE.

These are: (i) the loading of the negatively charged carrier with its substrate L-Lysine together with two hydroxyl ions, (ii) substrate translocation via the membrane, (iii) the release of L-Lysine and the accompanying ions at the outside of the membrane, and finally, (iv) the reorientation of the carrier. The driving force for the entire translocation process is the membrane potential, $\Delta\psi$, required for the reorientation of the carrier.

Access to the lysine-exporter gene, lysE, has also made it possible to solve the puzzle as to why c. glutamicum has such an exporter at all. In a lysE deletion mutant supplied with glucose and 1 mmol L^{-1} of the dipeptide, lysyl-alanine, an extraordinarily high intracellular L-Lysine concentration of more than 1 mol L^{-1} accumulates, abolishing growth of the mutant. Thus, the exporter serves as a valve to excrete any excess intracellular L-Lysine that may arise in the natural environment in the presence of peptides. As in the case of other bacteria, too, c. glutamicum has active peptide-uptake systems as well as hydrolysing enzymes giving access to the amino acids as valuable building blocks. However, c. glutamicum has no L-Lysine-degrading activities and therefore must prevent any piling up of L-Lysine. This also happens in the lysine producer strains where the biosynthesis pathway is mutated. As genome projects have now shown, homologous structures of the L-Lysine carrier LysE are present in various Gram-negative and Gram-positive bacteria. Therefore, this type of intracellular amino acid control by an exporter is expected to be present in other bacteria, too. Since the LysE structure is not shared with other translocators, LysE also represents a new superfamily of translocators, which is probably related to its new function.

5 Production strains

L-Lysine producer strains have been derived over the decades by mutagenesis to give strains excreting more than 170 g L-Lysine per litre. It is clear that these strains carry a long list of phenotypic characters to achieve this massive flux directioning (Tab. 1.2). Typically, the strains are resistant or sensitive to some analogue of lysine. A typical feature of some L-Lysine producers is their resistance to the lysine analogue S-(2-aminoethyl)-L-Cysteine. In these mutants, the aspartate kinase is mutated so that it is no longer inhibited by L-Lysine. Dozens of other chemicals structurally related to L-Lysine, such as γ-methyl-L-Lysine or α-chlorocaprolactam, have been used in screenings to obtain improved producers. Fluoropyruvate has also been used to identify strains that are sensitive to it as these have decreased pyruvate dehydrogenase activities resulting in a diminished oxidation of pyruvate via the citric acid cycle. This is also the case in strains with decreased citrate synthase activity. In another lineage of strains, over-producers were derived from mutants with diminished homoserine dehydrogenase activity to lower the availability of

L-Threonine inside the cell. In this way, inhibition of the kinase activity was abolished and, at the same time, a favourable growth limitation was introduced.

6 Production process

The most common carbon sources for L-Lysine fermentation and also other amino acids are molasses (cane or sugar beet molasses), high test molasses (inverted cane molasses) or sucrose and starch hydrolysates. In contrast to e. coli, the wild type of c. glutamicum can utilise both glucose and sucrose. There are also production technologies available based on acetic acid or ethanol as feedstocks. In the past, molasses was mostly used for production since it is a relative cheap carbon source. However, the utilization of molasses has severe disadvantages:

Waste is exported from the sugar company to the fermentation plant and causes additional costs there;

The seasonal availability of molasses causes ageing effects in its quality during storage.

Therefore, there is a clear tendency away from molasses towards refined carbon sources such as hydrolysed starches. Profitable nitrogen sources are ammonium sulphate and ammonia (gaseous or ammonia water). The growth factors required are provided from plant protein hydrolystates, corn steep liquor or by the addition of the defined compounds. A typical lysine fermentation is shown in Fig. 1.13. After consumption of the initial sugar, the substrates are added continuously and L-Lysine accumulates up to 170 g L^{-1}. Ammonium sulphate provides the counterion to neutralise the accumulating basic amino acid. Therefore, L-Lysine is present in the fermentation broth as its sulphate. As a convention in the literature, lysine is usually given as lysine HCl. Due to the high sugar cost the conversion yield is a very important criterion for the entire production process. Technical processes have been published with a yield of 45 – 50 g Lys. HCl per 100 g carbon source.

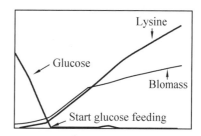

Fig. 1.13　Time course of L-Lysine accumulation in a production plant. There are three phases of growth and L-Lysine accumulation

For the recovery of L-Lysine, several basically different processes have been developed. Three processes are currently in use to supply L-Lysine in a form suitable for feed purposes:

A crystalline preparation containing 98.5% L-Lysine. HCl. It can be made by ion

exchange chromatography, evaporation and crystallisation. Also direct spray-drying of the ion exchange eluate is possible.

An alkaline solution of concentrated L-Lysine containing 50% L-Lysine. It is obtained by biomass separation, evaporation and filtration.

A granulated lysine sulphate preparation consisting of 47% L-Lysine. It consists of the entire fermentation broth conditioned by spraydrying and granulation.

These processes differ in investment costs, losses during downstreaming, amount of waste volume, and user friendliness. All this, together with the fermentation itself, decides the success of the entire production process.

3 L-Threonine

The commercial production of L-Threonine is possible with either e. coli or c. glutamicum mutants. However, the production figures of selected e. coli strains are superior. The synthesis of L-Threonine proceeds via a short pathway comprising only five steps. As already mentioned, the first steps are shared with that of L-Lysine and L-Methionine synthesis. Furthermore, L-Threonine is also an intermediate in the L-Isoleucine synthesis. This naturally requires special metabolic regulation. In c. glutamicum this was solved in such a way that the sole aspartate kinase present was only inhibited by the joint presence of L-Lysine and L-Threonine. In the case of e. coli, however, three isoenzymes are present each of which is separately inhibited by a different end-product: one by L-Threonine, one by L-Lysine and one by L-methionine. There are furthermore two homoserine dehydrogenase activities: one is inhibited by L-Threonine, and one by L-methionine. Additionally, the corresponding genes are grouped into transcriptional units, thereby ensuring a balanced synthesis of the appropriate amino acid at the level of gene expression. The relevant operon for L-Threonine synthesis in e. coli is thrABC. It encodes three polypeptides, with thrA encoding an apparently fused polypeptide with kinase plus dehydrogenase activity. Therefore, four enzyme activities of the five steps required to convert L-Aspartate to L-Threonine are encoded by thrABC. A strong expression control of this operon is provided by a transcription attenuation mechanism. The corresponding leader peptide at the beginning of the transcription unit is Thr-Thr-Ile-Thr-Thr-Thr-Ile-Thr-Ile-Thr-Thr, serving to sense the availability of L-Threonine and L-Isoleucine. When the corresponding tRNAs are uncharged, the leader peptide formation does not occur, and transcription of the operon is increased at least ten-fold.

Unit 1

1 Producer strains

Based on this regulation there is a clear focus on two major targets for the design of a producer strain: the prevention of L-Isoleucine formation, and stable high-level expression of thrABC. Therefore, in one of the first steps of strain development, chromosomal mutations were introduced to give an isoleucine leaky strain. The isoleucine mutation is a very specific and important one. L-Isoleucine is required for growth only at low L-Threonine concentrations but, at high concentrations of L-Threonine, growth is independent of L-Isoleucine. The mutation therefore has several advantageous consequences. In the first place, it prevents an excess formation of the undesired byproduct L-Isoleucine. Additionally, it prevents the L-Isoleucine dependent premature termination of the thrABC transcription due to limiting tRNA^{Ile}. A high transcription rate is, of course, required to have high specific enzyme activities.

Another consequence of the isoleucine mutation is more subtle. It relates to the stability of the plasmid-containing producer strain in the various pre-cultivation steps. Starting from a single clone, a preculture is inoculated for each production run and is then enlarged in several stages. This means that the clone is fermented for about 25 generations so that there is a great danger of the plasmid containing the thrABC operon being lost. This would of course be a complete disaster if it happened in the final production stage. In the presence of the isoleucine leaky mutation, however, cells that have lost the plasmid now are clearly disadvantaged when not supplied with L-Isoleucine. Their further proliferation is halted, thereby stabilising a culture where almost all the cells that are growing contain the plasmid. Further engineering during strain evolution involved the introduction of resistance to L-Threonine and L-Homoserine. Subsequently, tdh, which encodes threonine dehydrogenase, was inactivated thus preventing threonine degradation. To obtain very high activities of the thrABC-encoding enzymes, the operon was cloned from a strain whose kinase and dehydrogenase activities are resistant to L-Threonine inhibition. In addition, the transcription attenuator region was deleted. In fermentations the operon engineered in this way was successfully used with pBR322 as a vector, but a further improvement was obtained by replacing this plasmid by a pRS1010 derivative, resulting in an even more stable high-level expression.

2 Substrate uptake

Since the cost of the sugar source has a decisive influence on the price of the amino acid produced it is essential to be able to switch between glucose and sucrose as substrates. However, only a few of the e. coli strains can use sucrose. Two different sucrose-utilising systems of e. coli are available to engineer sugar utilisation in L-Threonine producing strains. One of them is represented by the scr regulon, where the actual translocator consists of a phosphoenol pyruvate: sugar phosphotransferase system (PTS). Introduction of the scr genes into a glucose-utilising e. coli strain results in the uptake and phosphorylation of sucrose. Due to subsequent hydrolase and fructokinase activities the sugar is then channelled into the central metabolism. An alternative sucrose utilisation system is provided by the csc regulon of some e. coli strains. In this case, sucrose is translocated by the cscB encoded translocator in symport with protons. Using transposition the sucrose-utilisation capability of the csc regulon was introduced into a glucose-utilising strain. Although originally without uptake of sucrose, this strain now imported sucrose at a rate of 9 pmol $min^{-1} \cdot$ mg dry wt. With the plasmid-encoded regulon, the rate obtained was 43 pmol $min^{-1} \cdot$ mg cell dry wt, which was almost identical to that of the strain from which the csc regulon had been isolated.

3 Production process

The fermentation of the engineered L-Threonine producer is in a simple mineral salts medium with either glucose or sucrose as the substrate with addition of a small amount of a complex medium component like yeast extract. After the inoculation and consumption of the initially provided sugar, continuous feeding of sugar begins. Additionally, ammonia has to be fed in the form of gas or as NH_4OH, which is regulated via pH control. Thus the feeding strategy in the case of L-Threonine fomentation is quite easy compared to L-Lysine fermentation where the accumulation of the basic product requires the feeding of sulphate as the counter-ion. At the end of the fermentation, L-Threonine is present in concentrations of about 85 g L^{-1} with a conversion yield of up to 60% based on the carbon source used. Such fermentations with high yields show quite low byproduct levels. This is an advantage for downstream processing. Crystallisation of L-Threonine is easy due to its low solubility (about 90 g L^{-1} in water) and the low salt concentration present. A process is described where the cells are initially coagulated by a heat—or pH-treatment step, followed by filtration. Subsequently, the broth is concentrated and crystallisation initiated by cooling. The separation and drying of the crystals leads to an isolation yield of 80% to 90% with the L-Threonine having a purity of more than 90%. A recrystallisation step may be required for high-purity L-Threonine.

Unit 1

4 L-Phenylalanine

L-Phenylalanine can be produced with e. coli or c. glutamicum. The pathway for L-Phenylalanine synthesis is shared in part with that of L-Tyrosine and L-Tryptophan. These three aromatic amino acids have in common the condensation of erythrose 4-phosphate and phosphoenol pyruvate to deoxyarabinoheptulosonate phosphate (DAHP) with further conversion in six steps up to chorismate. L-Phenylalanine is then finally made in three further steps. There are three DAHP synthase enzymes in e. coli encoded by aroF, aroG and aroH. These enzymes play a key role in flux control. Their regulation of catalytic activity, in each case by one of the three aromatic amino acids, recalls the specific regulation of aspartate kinase in the synthesis of threonine. About 80% of the total DAHP-synthase activity is contributed by the aroG-encoded enzyme. Increased flux towards L-Phenylalanine can be obtained by over-expression of either aroF or aroG encoding feedback-resistant enzymes. Furthermore, pheA over-expression is essential. This gene encodes the bifunctional corismate mutase-prephenate dehydratase. A second chorismate activity is present as a bifunctional chorismate mutase-prephenate dehydrogenase. The pheA-encoded enzyme activities are inhibited by L-Phenylalanine and pheA expression is dependent on the level of $tRNA^{phe}$.

1 Production strains

Producer strains have a DAHP activity that is resistant to feedback inhibition and which is encoded either by aroF or aroG and a feedback-resistant corismate mutase-prephenate dehydratase. As a rule, the producers are L-Tyrosine auxotrophic mutants. There are very good reasons for this, one of which is that the enzymes of the common pathway from DAHP to prephenate are no longer regulated by L-Tyrosine and enzyme activities are no longer feedback-inhibited. Another reason is that in this way tyrosine accumulation is prevented, which would otherwise undoubtedly result as a byproduct since there are only two additional steps from prephenate to L-Tyrosine. An essential aspect is that due to the auxotrophy, a beneficial growth limitation is possible by appropriate tyrosine feeding. In some e. coli strains, the temperature-sensitive cI_{857} repressor of bacteriophage λ has been used together with the λP_L promoter to enable inducible expression of the key genes pheA and aro F. This enables extremely high enzyme activities to be adjusted solely in the actual production runs thus eliminating the inherent problems of strain stability due to the resulting high metabolite concentrations or side activities of the enzymes. It enables the pre-cultivation steps up to the seed fermenter to be performed with low expression of the key genes but in the actual large production fermenter the genes are now induced to a high level of expression.

2 Production process

As with the other amino acids, effective L-Phenylalanine production is the joint result of engineering the cellular metabolism and control of the production process. Control is necessary for two reasons. First, the carbon flux has to be optimally distributed between the four major products of glucose conversion, which are L-Phenylalanine, biomass, acetic acid and CO_2. The second reason is that the cellular physiology is not constant during the course of fermentation, which correspondingly requires an adaptation of fermentation control during the process. Fig. 1.14 shows the typical time curve of L-Phenylalanine production. The major problem is that e. coli tends to produce acetic acid, which has a strong negative effect on process efficiency. To prevent this, researchers have developed an ingenious sugar-feeding strategy, which first collects on-line data and fluxes such as oxygen concentration, sugar consumption and biomass concentrations. These are then counterbalanced during

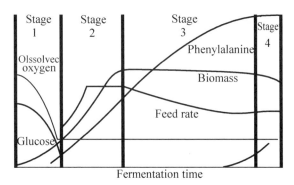

Fig. 1.14 The four stages of L-Phenylalanine production characterized by different physiology requiring different process control regimes to give the highest yields in shortest times

the process to control the optimum sugar concentration. The feeding of sugar starts when the cells enter Stage 2 of the fermentation where the glucose initially provided has almost been consumed. The trick is to prevent too high a glucose concentration occurring since this would result in acetic acid formation and, at the same time, to prevent too low a glucose concentration since this would result in an excess of CO_2 evolution. Thus the feeding rate is a compromise where the process is run at the highest possible feeding rate, which still provides a sufficiently strong limitation to prevent acetic acid excretion. When the L-Tyrosine initially present has been consumed, the cells proceed to Stage 3. As already mentioned, almost all L-Phenylalanine producers cannot synthesise tyrosine. The L-Tyrosine concentration selected at the start of the culture therefore fixes the minimum amount of biomass necessary to efficiently metabolise the predetermined amount of glucose. In Stage 3, the metabolic capacity of the cells decreases which brings about a consequent decrease of the glucose feeding rate. At the end of Stage 3, acetic acid excretion begins and the cells enter Stage 4 where no further L-Phenylalanine accumulation occurs and the process is eventually terminated. This example of amino acid production shows that by the sophisticated application of feeding strategies with adaptive control a very high L-Phenylalanine concentration can be achieved with a high yield within 2.5 days. Values of 50.8 g phenylalanine per litre with a yield of 27.5% of carbon used have been reported.

5 L-Tryptophan

L-Tryptophan is a high-price amino acid which still has a rather low market volume. Effective production processes are available with mutants of different bacteria, including Bacillus subtilis. However, cellular synthesis is no longer performed due to originally not realized impurities in the final product used for medical purposes. These impurities arose during the isolation of L-Tryptophan from a chemical reaction with traces of acetaldehyde at low pH. An alternative process is the enzymatic synthesis of L-Tryptophan from precursors. The current enzymatic production process uses the activity of the biosynthetic tryptophan synthase. This enzyme catalyses the last step in the tryptophan synthesis, which consists in fact of two partial reactions:

$$\text{Indole 3 - glycerol phosphate} \rightarrow \text{indole} + \text{glyceraldehyde 3 - phosphate}$$
$$\text{Indole} + \text{L-serine} \rightarrow \text{L-Tryptophan} + H_2O$$

These separate reactions are catalysed by separate subunits of the enzyme: α and β. The enzyme of e. coli is an $\alpha_2\beta_2$ tetramer, which can be dissociated into two α subunits and a β_2 subunit. The α subunit catalyses the cleavage of indole 3 - glycerol phosphate, whereas the β_2 subunit catalyses the condensation of L-Serine with indole to form L-Tryptophan. Each β-subunit contains one molecule of covalently bound pyridoxal phosphate, forming a Schiff's base with L-Serine. This enzyme-bound aminoacrylate is attacked when indole is provided from the α subunit. But how does indole get to the β-subunit? The problem is that indole is very hydrophobic so that with free diffusion it can pass through the cell membrane and be lost. The crystal structure of the synthase revealed the ingenious solution for solving this problem. To prevent a loss of indole it is channelled within the enzyme protein. There is a 25 nm long tunnel from the α subunit, where indole is formed, to the β-subunit where, as the enzyme-bound aminoacrylate, L-Serine is ready to accept the indole. Furthermore, within the native tetramer both partial reactions are coordinated. Only when L-Serine, as aminoacrylate, is ready to accept indole, does indole 3-glycerol phosphate conversion occur at the β-subunit. Tryptophan synthase is thus an example of how an enzyme complex is used as a sophisticated device to handle a reactive and diffusible intermediate within the cell.

1 Production from precursors

The process of L-Tryptophan production with this enzyme is based on e. coli cells which have a high tryptophan synthase activity. The α and β-subunits encoding genes trpA and trpB, respectively, are located on the trpEDCBA operon, which is regulated, by repression and attenuation. In the e. coli mutant used, the represser of that operon has been deleted as is part of the attenuator region together with the first structural genes of the operon. In the resulting strain, about 10% of the total protein is tryptophan synthase with an excess of the β-subunit. Although indole is not the true substrate of the enzyme, with a sufficiently high concentration the enzyme will react with it. Indole is available from the petrochemical industry as a comparably cheap educt, whereas the second educt, L-Serine, is recovered from molasses during sugar refinement using ion exclusion chromatography, and further purification steps. The resulting L-Serine is fed to the previously cultivated e. coli cells, and indole is added continuously at a concentration adjusted to 10 mmol L^{-1}, which is controlled on-line. This type of process ensures an almost quantitative conversion of indole to yield L-Tryptophan with a space-time yield of about 75 g per litre and day.

6 L-Aspartate

L-Aspartic acid is widely used as a food additive and in pharmaceuticals, Demand increased rapidly with the introduction of aspartame as an artificial sweetener. This is a dipeptide consisting of L-Aspartate and L-Phenylalanine, which is about 200-fold sweeter than sugar and was successfully introduced into the market as a low-calorie sweetener. Although L-Aspartate was originally produced fermentatively, it is currently produced exclusively using aspartase due to the high productivities and the cost effectiveness of the process. In fact, the use of aspartase to make L-Aspartate represents one of the highest productivities known for an enzyme used in biotechnology. The method developed allows reuse of the enzyme to the extent that over 220,000 kg of product can be produced per kg of enzyme.

Aspartase catalyses the interconversion between L-Aspartate and fumarate plus ammonia. The reaction favours the animation reaction. The enzyme of e. coli is a tetramer with a molecular weight of 196,000 which has an absolute requirement for divalent metal ions. A severe disadvantage at the beginning of the work by the Tanabe Seiyaku Company, which now successfully uses aspartase, was the instability of the enzyme. After incubation of the enzyme in solution for just half an hour at 50 ℃, activity was no longer detectable. Nevertheless, a residual activity of 10% is present when the enzyme is immobilized in polyacrylamide. Such a physical confinement of cells in space turned out to be the method of choice. With the natural polymer K-carrageenan, resulting from a screening of different polymers, and use of appropriate cross-linking exceptional improvements are obtained in the relative productivity as well as in the stability of the catalyst. The final material has a half-life of almost two years. This represents almost unimaginable progress in comparison to the initial situation where enzyme in free solution only had a half-life measured in minutes. An initial disadvantage of the original cells used, was their fumarase activity which results in the partial conversion of fumarate to L-Malic acid. To solve this problem a heat treatment step of the cells is used which eliminates the fumarase activity almost completely. Using such conditioned cells and starting with 1 mol L^{-1} ammonium fumarate, the final product solution contains 987 mmol L^{-1} L-Aspartate, 10.7 mmol L^{-1} non-reacted fumarate and only trace quantities of L-Malic acid of 1.9 mmol L^{-1}.

For the production process the immobilized cells are packed into a column designed as a multistage system. The stages introduced, consisting of horizontal tubes, serve two purposes. On the one hand, they allow effective cooling to prevent decay of the catalytic activity since the aspartase reaction is exergonic. About 6 kcal heat mol^{-1} substrate evolves in the actual large-scale production process which is very close to that calculated from the standard free energy change of the aspartase reaction of 4 kcal mol^{-1}. On the other hand, the flow properties of the column are increased. Any compacting of the bed over time is prevented, and the preferred plug-flow characteristics are obtained. With such a column, flow rates of two column volumes per hour are possible. The continuous process enables full automation and control to achieve an optimum throughput with the highest product quality. Yet another advantage of such a controlled continuous process is its reduced waste production. A typical volumetric activity is about 200 mmol h^{-1} g cells. Assuming a 1,000 litre column, the yield of L-Aspartate is 3.4 tonnes per day which is 100 tonnes per month. The final product is eventually purified by crystallisation.

Unit 2

··· Part A ···

Brief History of Enzymology

 Introduction

Life depends on a well-orchestrated series of chemical reactions. Many of these reactions, however, proceed too slowly on their own to sustain life. Hence nature has designed catalysts, which we now refer to as **enzymes**, to greatly accelerate the rates of these chemical reactions. The catalytic power of enzymes facilitates life processes in essentially all life-forms from **viruses** to man. Many enzymes retain their catalytic potential after extraction from the living organism, and it did not take long for mankind to recognize and exploit the catalytic power of enzyme for commercial purposes. In fact, the earliest known references to enzymes are from ancient texts dealing with the manufacture of **cheeses**, breads, and **alcoholic beverages**, and for the **tenderizing** of meats. Today enzymes continue to play key roles in many food and beverage manufacturing processes and are ingredients in numerous consumer products, such as laundry **detergents** (which dissolve **protein**-based stains with the help of proteolytic enzymes). Enzymes are also of fundamental interest in the health sciences, since many disease processes can be linked to the aberrant activities of one or a few enzymes. Hence, much of modern **pharmaceutical** research is based on the search for potent and specific **inhibitors** of these enzymes. The study of enzymes and the action of enzymes has thus fascinated scientists since the dawn of history, not only to satisfy **erudite** interest but also because of the utility of such knowledge for many practical needs of society. This brief chapter sets the stage for our studies of these remarkable **catalysts** by providing

Unit 2

a historic background of the development of enzymology as a science. We shall see that while enzymes are today the focus of basic academic research, much of the early history of enzymology is linked to the **practical application** of enzyme activity in industry.

New Words

enzymology	酶学
enzyme	酶
virus	病毒,滤过性微生物
cheese	干酪
alcoholic beverage	酒精饮料
tenderize	使嫩,使肉变软
detergent	清洁剂,去垢剂
protein	蛋白质,蛋白质的
pharmaceutical	药物;制药(学)上的
inhibitor	抑制剂
erudite	博学的
catalyst	催化剂
practical application	实际应用

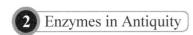

2 Enzymes in Antiquity

The oldest known reference to the commercial use of enzymes comes from a description of wine making in the **Codex of Hammurabi** (ancient **Babylon**, circa 2100 B. C.). The use of **microorganisms** as enzyme sources for **fermentation** was widespread among ancient people. References to these processes can be found in writings not only from Babylon but also from the early civilizations of Rome, Greece, Egypt, China, India. Ancient texts also contain a number of references to the related process of vinegar production, which is based on the **enzymatic conversion** of **alcohol** to **acetic acid. Vinegar**, it appears, was a common staple of ancient life, being used not only for food storage and preparation but also for medicinal purposes.

Dairy products were another important food source in ancient societies. Because in those days **fresh** milk could not be stored for any reasonable length of time, the conversion of milk to cheese became a vital part of food production, making it possible for the farmer to bring his product to distant markets in an acceptable form. Cheese is prepared by **curdling** milk via the action of any of a number of enzymes. The substances most commonly used for this purpose in ancient times were **ficin**, obtained as an extract from **fig trees**, and **rennin**, as **rennet**, an extract of the lining of the **fourth stomach** of a multiple-stomach animal, such as a cow. A reference to the enzymatic activity of ficin can, in fact, be found in Homer's classic, the Iliad:

As the juice of the fig tree curdles milk, and thickens it in a moment though it be liquid, even so instantly did Paion cure fierce Mars.

The philosopher Aristotle likewise wrote several times about the process of milk curdling and offered the following **hypothesis** for the action of rennet:

Rennet is a sort of milk; it is formed in the stomach of young animals while still being suckled. Rennet is thus milk which contains fire, which comes from the heat of the animal while the milk is undergoing concoction.

Another food staple throughout the ages is bread. The **leavening** of bread by yeast, which results from the enzymatic production of **carbon dioxide**, was well known and widely used in ancient times. The importance of this process to ancient society can hardly be overstated.

Meat tenderizing is another enzyme-based process that has been used since antiquity. Inhabitants of many Pacific islands have known for centuries that the **juice** of the **papaya** fruit will soften even the toughest meats. The active enzyme in this plant extract is a **protease** known as **papain**, which is used even today in commercial meat tenderizers. When the British Navy began exploring the Pacific islands in the 1700s, they encountered the use of the papaya fruit as a meat tenderizer and as a treatment for **ringworm**. Reports of these native uses of the papaya sparked a great deal of interest in eighteenth-century Europe, and may, in part, have led to some of the more **systematic** studies of digestive enzymes that ensued soon after.

New Words

antiquity	古代, 古老
Codex of Hammurabi	汉穆拉比法典
Babylon	巴比伦
microorganism	微生物, 微小动植物
fermentation	发酵

enzymatic conversion	(淀粉)酶糖化;(蔗糖)酶转化
alcohol	酒精,酒
acetic acid	乙酸,醋酸
vinegar	醋
dairy products	乳制品
fresh	新鲜的
curdling	凝化
ficin	无花果蛋白酶
fig tree	无花果树
rennin	高血压蛋白原酶
rennet	凝乳酶
fourth stomach	第四胃,皱胃
Iliad	《伊利亚特》(古希腊描写特洛伊战争史诗)
hypothesis	假设
suckle	哺乳,养育,吮吸
concoction	调和,混合,调和物
leavening	酵母,发酵,发酵物
carbon dioxide	氧化碳
juice	(水果)汁,液
papaya	木瓜
protease	蛋白酶
papain	木瓜蛋白酶
ringworm	癣,癣菌病
systematic	系统的,体系的

 Early Enzymology

While the ancients made much practical use of enzymatic activity, these early applications were based purely on **empirical** observations and folklore, rather than any systematic studies or appreciation for the chemical basis of the processes being utilized. In the eighteenth and nineteenth centuries scientists began to study the actions of enzymes in a more systematic fashion. The process of digestion seems to have been a popular subject of investigation during the years of the enlightenment. Wondering how **predatory birds** manage to digest meat without a **gizzard**, the famous French scientist **Réaumur** (1683—1757) performed some of the earliest studies on the digestion of **buzzards**. Réaumur designed a metal tube with a wire mesh at one end that would hold a small piece of meat immobilized, to protect it from the physical action of the **stomach** tissue. He found that when a tube containing meat was inserted into the stomach of a buzzard, the meat was digested within 24 hours. Thus he concluded that digestion must be a chemical rather than a merely physical process, since the meat in the tube had been digested by contact with the **gastric juices** (or, as he referred to them, "a solvent"). He tried the same experiment with a piece of bone and with a piece of a plant. He found that while meat was digested, and the bone was greatly softened by the action of the gastric juices, the plant material was impervious to the "solvent"; this was probably the first experimental **demonstration** of enzyme specificity.

Réaumur's work was expanded by **Spallanzani** (1729—1799), who showed that the digestion of meat encased in a metal tube took place in the stomachs of a wide variety of animals, including humans. Using his own gastric juices, Spallanzani was able to perform digestion experiments on pieces of meat **in vitro** (in the laboratory). These experiments illustrated some critical features of the **active ingredient** of gastric juices: by means of a control experiment in which meat treated with an equal volume of water did not undergo digestion Spallanzani demonstrated the presence of a specific active ingredient in gastric juices. He also showed that the process of digestion is temperature dependent, and that the time required for digestion is related to the amount of gastric juices applied to the meat. Finally, he demonstrated that the active ingredient in gastric juices is unstable outside the body; that is, its ability to digest meat wanes with storage time.

Today we recognize all the foregoing properties as common features of enzymatic reactions, but in Spallanzani's day these were novel and exciting findings. The same time period saw the discovery of enzyme activities in a large number of other biological systems. For example, a **peroxidase** from the **horseradish** was described, and the action of α-**amylase** in grain was observed. These early observations all pertained to materials—crude extract from plants or animals—that contained enzymatic activity.

During the latter part of the nineteenth century scientists began to attempt **fractionations** of these extracts to obtain the active ingredients in pure form. For example, in 1897 Bertrand partially purified the enzyme **laccase** from tree **sap**, and Buchner, using the "**pressed juice**" from **rehydrated** dried yeast, demonstrated that alcoholic fermentation could be performed in the absence of living yeast cells. Buchner's report contained the interesting observation that the activity of the pressed juice diminished within 5 days of storage at ice temperatures. However, if the juice

Unit 2

was supplemented with cane sugar, the activity remained intact for up to 2 weeks in the ice box. This is probably the first report of a now well-known phenomenon—the **stabilization** of enzymes by **substrate**. It was also during this period that Kühne, studying catalysis in yeast extracts, first coined the term "enzyme" (the word derives from the medieval Greek word **enzymos**, which relates to the process of leavening bread).

New Words

empirical	完全根据经验的,实验式
the enlightenment	(18世纪欧洲的)启蒙运动
predatory bird	捕食鸟
gizzard	(鸟的)砂囊,(鸟的)胃,(人的)胃
Réaumur	列奥米尔(1683—1757)法国物理学家,发明了酒精温度计,并设立创造了列氏温标
buzzard	秃鹰类
stomach	胃,胃口,胃部
gastric juice	胃液
demonstration	示范,实证
Spallanzani	斯布莱兹尼(生理学家)
in vitro	在试管中,在生物体外
active ingredient	活性组分,有效成分
preoxidase	过氧化酶
horseradish	辣根(一种十字花科的粗糙植物)
amylase	淀粉酶
fractionation	分馏,分馏法
laccase	漆酶,虫漆酶
sap	树液,体液
press juice	榨出汁,滤液
rehydrate	再水化,再水合,使复水
stabilization	稳定性
substrate	培养基,酶作用物,底物
enzymos	指在别的事物中起激励作用

4 The Development of Mechanistic Enzymology

As enzymes became available in pure, or partially pure forms, scientists' attention turned to obtaining a better understanding of the details of the **reaction mechanisms** catalyzed by enzymes. The concept that enzymes form complexes with their substrate molecules was first articulated in the late nineteenth century. It is during this time period that **Emil Fischer** proposed the "lock and key" model for the stereochemical relationship between enzymes and their substrates; this model emerged as a result of a large body of **experimental** data on the **stereospecificity** of enzyme reactions. In the early twentieth century, experimental evidence for the formation of an enzyme-substrate complex as a reaction intermediate was reported. One of the earliest of these studies, reported by Brown in 1902, focused on the velocity of enzyme-catalyzed reactions. Brown made the insightful observation that unlike simple **diffusion**-limited chemical reactions, in enzyme-catalyzed reactions "it is quite conceivable... that the time elapsing during molecular union and transformation may be sufficiently prolonged to influence the general course of the action." Brown then went on to summarize the **available data** that supported the concept of formation of an enzyme-**substrate complex**:

There is reason to believe that during inversion of cane sugar by invertase the sugar combines with the enzyme previous to inversion. C. O'Sullivan and Tompson... have shown that the activity of invertase in the presence of cane sugar survives a temperature which completely destroys it if cane sugar is not present, and regard this as indicating the existence of a combination of the enzyme and sugar molecules. Wurtz [1880] has shown that papain appears to form an insoluble compound with fibrin previous to hydrolysis. Moreover, the more recent conception of E. Fischer with regard to enzyme configuration and action, also implies some form of combination of enzyme and reacting substrate.

Observations like these set the stage for the derivation of enzyme rate equations, by mathematically modeling enzyme **kinetics** with the explicit involvement of an **intermediate** enzyme-substrate complex. In 1903 Victor Henri published the first successful mathematical model for describing enzyme kinetics. In 1913, in a much more widely read paper, Michaelis and Menten expanded on the earlier work of Henri and rederived the enzyme rate equation that today bears their names. The Michaelis-Menten equation, or more correctly the Henri-Michaelis-Menten equation, is a **cornerstone** of much of the modern analysis of enzyme reaction mechanisms.

Unit 2

The question of how enzymes accelerate the rates of chemical reactions puzzled scientists until the development of **transition state theory** in the first half of the twentieth century. In 1948 the famous physical chemist Linus Pauling suggested that enzymatic rate **enhancement** was achieved by stabilization of the transition state of the chemical reaction by interaction with the enzyme active site. This **hypothesis**, which was widely accepted, is supported by the experimental observation that enzymes bind very tightly to molecules designed to **mimic** the structure of the transition state of the catalyzed reaction.

In the 1950s and 1960s scientists **reexamined** the question of how enzymes achieve substrate specificity in light of the need for transition state stabilization by the enzyme active site. New hypotheses, such as the "**induced fit**" **model** of Koshland emerged at this time to help rationalize the competing needs of substrate binding **affinity** and reaction rate enhancement by enzymes. During this time period, scientists struggled to understand the observation that metabolic enzyme activities can be regulated by small molecules other than the substrates or direct products of an enzyme. Studies showed that indirect interactions between distinct binding sites within an enzyme molecule could occur, even though these binding sites were quite distant from one another. In 1965 Monod, Wyman, and Changeux developed the theory of allosteric transitions to explain these observations. Thanks in large part to this **landmark** paper, we now know that many enzymes, and **nonenzymatic ligand** binding proteins, display **allosteric** regulation.

New Words

reaction mechanism	反应机理/机制
Emil Fischer	生物化学的创始人费歇尔
"lock and key" model	锁与钥匙模型
stereo	立体的
experimental data	实验数据
stereospecificity	(立体)定向性
diffusion	扩散,传播
available data	现有数据,现有资料
substrate complex	底物复合物
invertase	转化酵素
molecule	分子

cane sugar	蔗糖
papain	木瓜蛋白酶
fibrin	(血)纤维蛋白,(血)纤维
hydrolysis	水解
configuration	构造,结构
kinetics	动力学
intermediate	中间的;媒介
cornerstone	墙角石,基础
transition state theory	过渡态理论
enhancement	增进,增加
hypothesis	假设
mimic	模仿的,假装的,拟态的
reexamine	回顾
"induced fit" model	"诱导契合"模型
affinity	亲和力
landmark	里程碑,划时代的事
nonenzymatic	非酶的,不涉及酶作用的
ligand	配合基[体],向心配合(价)体
allosteric	变构(象)的

5 Studies of Enzyme Structure

One of the tenets of modern enzymology is that catalysis is intimately related to the molecular interactions that take place between a substrate molecule and components of the enzyme molecule, the exact nature and sequence of these interactions defining **perse** the catalytic mechanism. Hence, the application of physical methods to **elucidate** the structures of enzymes has had a rich history and continues to be of **paramount** importance today. **Spectroscopic** methods, **X-ray crystallography**, and more recently, **multidimensional NMR** methods have all provided a wealth of structural insights on which theories of enzyme mechanisms have been built. In the early part of the twentieth century, X-ray crystallography became the premier method for solving the structures of small molecules. In 1926 James Sumner published the first **crystallization** of an enzyme, **urease** (Fig. 2.1). Sumner's paper was a landmark contribution, not only because it portended the successful application of X-ray **diffraction** for solving enzyme structures, but also because a detailed analysis allowed Sumner to show **unequivocally** that the crystals were composed of protein and that their **dissolution** in solvent led to enzymatic activity. These observations were very important to the development of the science of enzymology because they firmly established the protein composition of enzymes, a view that had not been widely accepted by Sumner's contemporaries.

Sumner's crystallization of urease opened a **floodgate** and was quickly followed by reports of numerous other enzyme crystals. Within 20 years of Sumner's first paper more than 130 enzyme crystals had been documented. It was not, however, until the late 1950s that protein structures began to be solved through X-ray crystallography. In 1957 **Kendrew** became the first to deduce from X-ray diffraction the entire **three-dimensional** structure of a protein, **myoglobin**. Soon after, the crystal structures of many proteins, including

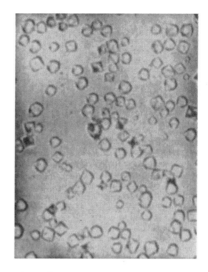

Fig. 2.1　Photomicrograph of urease crystals (728 × magnification), the first reported crystals of an enzyme

enzymes, were solved by these methods. Today, the structural insights gained from X-ray crystallography and multidimensional NMR studies are commonly used to elucidate the mechanistic details of enzyme catalysis, and to design new ligands (substrate and inhibitor molecules) to bind at specific sites within the enzyme molecule.

The deduction of three-dimensional structures from X-ray diffraction or NMR methods depends on knowledge of the **arrangement** of **amino acids** along the polypeptide chain of the protein; this arrangement is known as the amino acid sequence. To determine the amino acid sequence of a protein, the component amino acids must be **hydrolyzed** in a sequential fashion from the polypeptide chain and identified by chemical or **chromatographic analysis**. Edman and coworkers developed a method for the sequential hydrolysis of amino acids from the *N*-**terminus** of a **polypeptide chain**. In 1957 Sanger reported the first complete amino acid sequence of a protein, the **hormone insulin**, utilizing the chemistry developed by Edman. In 1963 the first amino acid sequence of an enzyme, **ribonuclease**, was reported.

New Words

perse	本身,本质上
elucidate	阐明,说明
paramount	极为重要的
spectroscopic	分光镜的,借助分光镜的
X-ray	X射线
crystallography	结晶学
multidimensional	多面的,多维的
NMR	核磁共振 nuclear magnetic resonance
crystallization	结晶化
urease	脲酶,尿素酶
diffraction	衍射
unequivocal	明确的
dissolution	分解
floodgate	水门,(河流,运河的)水闸
Kendrew	肯德鲁(英国生物学家)
three-dimensional	三维的,三度的,立体的
myoglobin	肌球素,肌血球素
arrangement	排列,安排
amino acid	氨基酸
hydrolyze	水解
chromatographic analysis	色谱(层)分析
N-terminus	N-末端
polypeptide chain	多肽链
hormone	荷尔蒙,激素
insulin	胰岛素
ribonuclease	核糖核酸酶

Unit 2

6 Enzymology Today

Fundamental questions still remain regarding the detailed mechanisms of enzyme activity and its relationship to enzyme structure. The two most powerful tools that have been brought to bear on these questions in modern times are the continued development and use of **biophysical probes** of **protein structure**, and the application of molecular biological methods to enzymology. X-ray crystallography continues to be used routinely to solve the structures of enzymes and of enzyme-ligand complexes. In addition, new NMR methods and **magnetization** transfer methods make possible the assessment of the three-dimensional structures of small enzymes in solution, and the structure of ligands bound to enzymes, respectively.

The application of **Laue** diffraction with **synchrotron radiation** sources holds the promise of allowing scientists to determine the structures of reaction intermediates during enzyme turnover, hence to develop detailed pictures of the individual steps in enzyme catalysis. Other biophysical methods, such as **optical** (e.g., **circular dichroism, UV-visible, fluorescence**) and vibrational (e.g., **infrared, Raman**) **spectroscopies**, have likewise been applied to questions of enzyme structure and reactivity in solution. Technical advances in many of these **spectroscopic** methods have made them extremely powerful and accessible tools for the enzymologist. Furthermore, the tools of molecular biology have allowed scientists to clone and express enzymes in **foreign host** organisms with great efficiency. Enzymes that had never before been isolated have been identified and characterized by molecular cloning. Over expression of enzymes in prokaryotic hosts has allowed the **purification** and **characterization** of enzymes that are available only in minute amounts from their natural sources. This has been a tremendous advance for protein science in general.

The tools of molecular biology also allow investigators to manipulate the amino acid sequence of an enzyme at will. The use of site-directed **mutagenesis** (in which one amino acid residue is substituted for another) and deletional mutagenesis (in which sections of the polypeptide chain of a protein are eliminated) have allowed enzymologists to pinpoint the chemical groups that participate in ligand binding and in specific chemical steps during enzyme catalysis.

The study of enzymes remains of great importance to the **scientific community** and to society in general. We continue to utilize enzymes in many industrial applications. Moreover enzymes are still in use in their traditional roles in food and beverage manufacturing. In modern times, the role of enzymes in consumer products and in chemical manufacturing has expanded greatly. Enzymes are used today in such varied applications as **stereospecific chemical synthesis**, laundry detergents, and cleaning kits for **contact lenses**.

Perhaps one of the most exciting fields of modern enzymology is the application of enzyme inhibitors as drugs in human and veterinary medicine. Many of the drugs that are commonly used today function by inhibiting specific enzymes that are associated with the disease process. **Aspirin**, for example, one of the most widely used drugs in the world, elicits its anti-**inflammatory** efficacy by acting as an inhibitor of the enzyme **prostaglandin synthase**. As illustrated in Tab. 2.1, enzymes take part in a wide range of human **pathophysiologies**, and many specific enzyme inhibitors have been developed to combat their activities, thus acting as **therapeutic** agents. Several of the inhibitors listed in Tab. 2.1 are the result of the combined use of biophysical methods for assessing enzyme structure and classical **pharmacology** in what is commonly referred to as rational or structure-based drug design. This approach uses the structural information obtained from X-ray crystallography or NMR spectroscopy to determine the **topology** of the enzyme **active site**. Next, model building is performed to design molecules that would fit well into this active site pocket. These molecules are then synthesized and tested as inhibitors. Several iterations of this procedure often lead to extremely potent inhibitors of the target enzyme.

The list in Tab. 2.1 will continue to grow as our understanding of disease state physiology increases. There remain thousands of enzymes involved in human physiology that have yet to be isolated or characterized. As more and more disease-related enzymes are discovered and characterized, new inhibitors will need to be designed to arrest the actions of these catalysts, in the continuing effort to fulfill unmet human medical needs.

Tab. 2.1 Examples of enzyme inhibitors as potential drugs

Inhibitor/Drug	Disease/Condition	Enzyme Target
Acetazolamide	Glaucoma	Carbonic anhydrase
Acyclovir	Herpes	Viral DNA polymerase
Allopurinol	Gout	Xanthine oxidase
Argatroban	Coagulation	Thrombin
Aspirin, ibuprofen, DuP697	Inflammation, pain, fever	Prostaglandin synthase
β-Lactam antibiotics	Bacterial infections	D-Ala-D-Ala transpeptidase
Brequinar	Organ transplantation	Dihydroorotate dehydrogenase

Tab. 2.1 (Continued)

Inhibitor/Drug	Disease/Condition	Enzyme Target
Candoxatril	Hypertension, congestive heart failure	Atriopeptidase
Captopril	Hypertension	Angiotensin-converting enzyme
Clavulanate	Bacterial resistance	β-Lactamase
Cyclosporin	Organ transplantation	Cyclophilin/calcineurin
DuP450	AIDS	HIV protease
Enoximone	Congestive heart failure ischemia	cAMP phosphodesterase
Finazteride	Benign prostate hyperplasia	Testosterone $-5-\alpha$-reductase
FK-506	Organ transplantation, autoimmune disease	FK-506 binding protein
Fluorouracyl	Cancer	Thymidilate synthase
3-Fluorovinylglycine	Bacterial infection	Alanine racemase
(2-Furyl)-acryloyl-Gly-Phe-Phe	Lung elastin degradation in cystic fibrosis	Pseudomonas elastase
ICI-200,808	Emphysema	Neutrophil elastase
Lovastatin	High cholesterol	HMG CoA reductase
Ly-256548	Inflammation	Phospholipase A_2
Methotrexate	Cancer	Dihydrofolate reductase
Nitecapone	Parkinson's disease	Catechol-O-methyltransferase
Norfloxacin	Urinary tract infections	DNA gyrase
Omeprazole	Peptic ulcers	$H^+, K^+-ATPase$
PALA	Cancer	Aspartate transcarbamoylase
PD-116124	Metabolism of antineoplastic drugs	Purine nucleoside phosphorylase
Phenelzine	Depression	Brain monoamine oxidase
Ro 42-5892	Hypertension	Renin
Sorbinil	Diabetic retinopathy	Aldose reductase
SQ-29072	Hypertension, congestive heart failure, analgesia	Enkephalinase

Tab. 2.1 (Continued)

Inhibitor/Drug	Disease/Condition	Enzyme Target
Sulfamethoxazole	Bacterial infection, malaria	Dihydropteorate synthase
Testolactone	Hormone-dependent tumors	Aromatase
Threo-5-fluoro-L-dihydroorotate	Cancer	Dihydroorotase
Trimethoprim	Bacterial infection	Dihydrofolate reductase
WIN 51711	Common cold	Rhinovirus coat protein
Zidovudine	AIDS	HIV reverse transcriptase
Zileuton	Allergy	5-Lipoxygenase

Source: Adapted and expanded from M. A. Navia and M. A. Murcko, Curr. Opin. Struct. Biol. 2, 202-210 (1992).

New Words

biophysical	生物物理的
probe	探针
protein structure	蛋白质结构
magnetization	磁性
Laue	劳厄(德国物理学家)
synchrotron radiation	同步加速器辐射
optical	光学的
circular dichroism	圆形(循环)二色性
UV-visible	紫外-可见的
fluorescence	荧光,荧光性
vibrational	振动的
infrared	红外的;红外线
Raman	拉曼(印度物理学家)
spectroscopy	光谱学,波谱学
spectroscopic	分光镜的,借助分光镜的
foreign	异质的,不相关的
host	宿主,寄主

Unit 2

purification	纯化
characterization	表征
mutagenesis	突变形成,变异发生
scientific community	科学界
stereospecific	立体定向的,立体专一的
chemical synthesis	化学合成(法)
contact lenses	隐形眼镜
aspirin	阿司匹林(解热镇痛药),乙酰水杨酸
inflammatory	炎性的,发炎的
prostaglandin	前列腺素
synthase	合酶
pathophysiology	病理生理学
therapeutic	治疗的,治疗学的
pharmacology	药理学
topology	拓扑,布局;拓扑学
active site	活性部位

Applications of Cellulase Enzymes

1 Stonewashing Denim

Denim stonewashing originated in the 1970s as a way to deliver pre-softened blue jeans to the public. The sewn denim was washed in the presence of pumice stones for roughly 60 minutes to shear and abrade the garments. The resulting jeans were softened by the stonewashing and therefore "ready to wear" at the time of purchase. This was preferred by the consumer over the previously-available stiff jeans that required an extensive "wear-in" period before they were comfortable. In addition, fashion favored faded jeans over the traditional dark blue jeans, and stonewashing removed a portion of the indigo dye, creating a faded appearance.

Although the marketplace favored stonewashed jeans, the use of stones caused several problems to denim washers. The stones damaged the washing machines, provided dust in the plant and the process effluent, and the handling of stones caused numerous worker injuries.

In the late 1980s, the use of cellulase enzymes began as an alternative to stones. Cellulase gained a foothold in the industry by producing the softness and appearance of jeans washed with pumice stones, but without the problems of stones. The availability of cellulase at costs competitive with stones in early 1990 led to the widespread adoption of cellulase use by the industry. Today, cellulase is used all over the world in this application, sometimes alone and sometimes with some stones present.

The original cellulases used in denim washing were the crude enzymes of Trichoderma and Humicola, referred to as "acid" and "neutral" cellulase, respectively, based on the optimum pH range of use of the enzymes, which was pH 4 to 5 for the acid cellulase and 6 to 7 for the neutral cellulase. The Trichoderma cellulase, comprising the more complete set of EG and CBH components capable of the full hydrolysis of cellulose, works more quickly and is capable of a greater degree of abrasion and fading of the blue dye color than the Humicola cellulase. The Trichoderma cellulase also achieves certain desired "finishes" (appearances) that the Humicola cellulase does not.

The Humicola cellulase works on the garments more gently, resulting in a higher retention of fabric strength. In addition, the Humicola cellulase results in less backstaining of the indigo dye onto the white threads and pockets of the denim. Finally, the neutral pH range of the Humicola cellulase facilitates easier and more forgiving pH adjustment of the wash, which is naturally slightly alkaline. For much of the market, the balance of advantages lay with the Humicola cellulase, which fostered a premium product status for it. However, the Trichoderma cellulase was still widely used.

By the mid-1990s, several new cellulase enzyme products had evolved for denim washing. Lower degrees of fabric strength loss were achieved with individual cellulase enzyme components, most prominently Trichoderma EG3 and Humicola EG5. Trichoderma cellulase used in combination with protease improved its performance to close to that of the Humicola cellulase. Some specialized cellulases are used to minimize dye streaking on the fabric, to cause more abrasion near seams, or any of a myriad of other desired effects. Cellulases operating at high pH (>9) have not been successful, as alkaline pH is used to terminate the enzyme reaction to prevent unwanted damage, and this procedure is ineffective with alkaline cellulase.

In terms of enzyme product formulations, liquid and granulated enzymes are used. Cellulase is combined with surfactants to give a cleaner finish.

❷ Household Laundry Detergent

The use of cellulase in household laundry detergent originated in 1993 with the introduction of Humicola EG5 into "New Cheer with Advanced Color Guard". Cellulase in laundry detergent removes the hairs, known as pills, that occur on cotton clothes after repeated wearing and machine washing. The cellulase removes the existing pills, and conditions the surface of new or unpilled clothes. The result is an appearance that more closely resembles a new garment in sharpness of color and smoothness of appearance. Cellulase also enhances the softness and removal of soil from the garment. The use of cellulase can eliminate the need for cationic fabric softeners, which have disposal and cost problems.

Household laundering is carried out at pH 8 to 9.5, and it is essential that the cellulase be active in this pH range. For this reason, the Trichoderma cellulases have not been successful in this application. The enzyme must also withstand potential inhibitors and inactivators, among them protease, surfactants, and bleach. The EG5 has been modified by protein engineering to improve its performance in detergent systems by increasing its tolerance of anionic surfactants, protease, and peroxide bleach and increasing its ability to adsorb to cellulose. Other cellulases used in household detergents include alkaline cellulase made by Bacillus, some of which is made in solid culture.

3 Animal Feed

Beta-glucanase enzymes were introduced to enhance the digestibility of animal feeds with limited success in the early 1970s. More recently, the cost-effective cellulase enzymes developed for other industries have been adapted for use in the feed industry.

The primary use of cellulase in the feed industry has been in barley- and wheat-based feeds for broiler chickens and pigs. The barley and wheat contain soluble beta-glucans that increase the viscosity of the feed in the gut of the animal. This, in turn, causes an uptake of water, which decreases the amount of carbohydrate and vitamins that the animal obtains from the feed, as well as causing sticky stool and related problems of disease and effluent disposal. Inclusion of cellulase in the feed, as well as xylanase and other enzymes, helps to overcome these problems.

This application is carried out with the crude cellulases of Trichoderma, Aspergillus, and Penicillium. The digestive process and action of cellulase in the system is not well understood. It seems to occur at acidic pH and crude cellulases from all three of these microbes are used.

The most important parameter characterizing feed digestibility is the feed conversion ratio (FCR), which is the ratio of weight of feed consumed to weight gain by the animal. Commercial cellulase enzymes, at dosages of 100 to 1,000 ppm in feed, decrease the FCR for barley and wheat feeds by 1% to 8% in field tests. This results is a significant cost savings to the farmers. The enzymes, which can be liquids or fine granules, are added directly to the feed just prior to consumption by the animals.

The use of cellulase in this application is most prominent in western Europe. The use of cellulase in corn/soy feeds, which are the most prominent in the US, has not been successful. Corn and soy are more easily digested than barley and wheat and less apt to result in a highly viscous feed.

Smaller applications of cellulase enzymes from submerged fermentation are in textile biopolishing, deinking and dewatering paper, processing of fruit juice and other beverages, baking, and alcohol production.

Unit 2

4 Textile Biopolishing

Biopolishing with cellulase is analogous to the use of cellulase in household laundry detergent. The enzyme removes pills from the fabric, restoring its appearance and conditioning it for resistance to further pilling.

In the present context, "biopolishing" is used to denote enzymatic depilling in an industrial context as carried out on unsewn fabric. There is a growing use of cellulase enzymes in the textiles industry.

Many fabrics, particularly blends of cotton and non-cotton fibers, develop a degree of hairiness during sewing or weaving. This hairiness is undesirable from the customers point of view, as it detracts from the overall appearance of the garment and the crispness of the colors. Lyocell, a blend of cotton and wood fibers, is particularly prone to pill formation.

The preferred method of removing the pills is to treat the pilled fabric with cellulase enzymes. This is typically carried out by loading the fabric onto a cylindrical drum that is partly submerged in liquor containing cellulase. The drum rotates at 50 – 100 rpm, which provides shear to help dislodge pills from the fabric. Cellulase treatment is typically for 15 – 30 min at 40 – 50 ℃, pH 5.

Most of the dyes used on Lyocell are more stable at acidic than alkaline pH, so this application has primarily been carried out using acidic cellulase from Trichoderma. The cellulase used can be the entire crude cellulase or modified cellulases with gentler action, depending on the size of the pills and the robustness of the fabric.

5 Deinking and Dewatering Paper

Two emerging applications of cellulase in the pulp and paper industry are in ink removal and water removal.

Deinking is the process by which the ink is removed from paper to allow it to be recycled. Deinking is carried out on low cost paper such as newsprint as well as fine paper, but cellulase is playing a role only in fine paper deinking at this time.

The conventional deinking process is by flotation. Fine paper, such as photocopier paper, is pulped in water at 5% to 20% solids consistency. The pulp is pumped to a flotation deinking basin, in which air is blown up through a sparger located near the bottom. The ink particles migrate to the air/liquid interfaces and float to the top of the basin. Surfactants are added to promote the migration of the ink particles. The ink is skimmed off the surface and the deinked pulp is bleached with hydrogen peroxide and sodium hydrosulfite, or in some cases more aggressive bleaching chemicals like chlorine dioxide or ozone.

Cellulase is added during pulping or flotation deinking. The enzyme increases the amount of ink removed from the fibers, thereby increasing the cleanliness of the sheet. This results in a brighter, cleaner sheet, or alternatively a reduction in the use of surfactants and bleaching chemicals.

The mechanism of the enzymes action in deinking is not well understood. The desired pH range for treating alkaline paper, which has become much more common than acid paper, is pH 7 to 8 to minimize degradation of the calcium carbonate. This is suited to the Humicola and Aspergillus cellulase enzymes, which are used in this application.

Paper dewatering is most important on the paper machine, where an aqueous slurry of pulp and additives are pressed into paper sheets. The water must be removed from the sheets by a combination of pressing, vacuum, and heat. The economy of the paper machine improves the faster it forms dry sheets of paper. However, speeding up the paper machine decreases the length of time available for dewatering. Often, the speed of the paper machine is limited by the rate at which water can be removed from the sheets.

Cellulase enzymes increase the rate of drainage of pulp, thereby offering the potential to increase the speed of the paper machine. Cellulase solubilizes a portion of the small particles, called fines, which are highly water holding. The rate of drainage is quantified by the Canadian Standard Freeness (CSF) measurement. Cellulase treatment increases the CSF by 20 to 40 points.

Cellulase treatment is carried out in a slurry tank prior to the paper machine. The typical treatment time is for 1 hour. Primarily Trichoderma cellulase enzymes have been used for this application, which is carried out at pH 4 to 5.

6 Fruit Juice and Beverage Processing

In the production of fruit juice, wine, beer, and other beverages, the raw juice is in a slurry (known as a "mash") with solid fruit. The juice is separated from the fruit by a combination of filtration and/or centrifugation. The separation operation is a cost trade-off: the more powder, time, and wash water used, the higher the yield of juice, but the more costly the process. There are also trade-offs of clarity and yield.

Cellulase enzymes break down cellulose and beta-glucan associated with the cell walls, thereby decreasing the viscosity of the mash and increasing the ease of the juice recovery. The conversion of cellulose and beta-glucans into soluble sugar provides another increase in the overall juice solids yield. The enzyme treatment can also increase the clarity of the juice by solubilizing small particles. The enzyme treatment can enhance the flavor of the juice by increasing the extractability of flavor compounds in the mash. Where disposal of the solid residue is costly, cellulase helps to decrease waste disposal costs.

Apple juice is the juice most commonly produced using cellulase enzymes, followed by cranberry juice, orange juice, and grape juice.

Although the benefits of cellulase are significant in juice production, pectinases are more important in these systems. Fruits contain high levels of pectin, and pectin is efficiently hydrolyzed by pectinases. Pectinase increases juice yields by 10% to 50%, while cellulase provides an enhancement of 5% to 15%. The effects of pectinase and cellulase together are additive and can even be synergistic.

Another concern in using cellulase in beverages, particularly beer and wine, is the possibility of changes in flavor. Although increasing flavor extractability is often desirable, many beer and wine brand names maintain a constant flavor that is undesirable to change.

Most cellulase used in the juice industry is Trichoderma cellulase, because of the typically low pH present in the mash. Cellulase from Aspergillus niger is also used.

7 Baking

Cellulase enzymes are used extensively in baking, but most of these are from solid culture. In general, the enzyme action desired in baking is very mild. Cellulase is used to break down gums in the dough structure, so as to allow a more even dough rise and flavor distribution. However, too much action can damage the dough structure and degrade the baked goods. Trichoderma cellulase, for example, is too aggressive in its hydrolytic action for cookie production, and its use in baking is restricted to crackers.

Among cellulases from submerged culture, Aspergillus is the most widely used in baking. Aspergillus cellulase is prominent where cellulase is used for baking breads and cakes. This stems from the use of Aspergillus cellulase from solid culture in baking.

8 Alcohol Production

Although not yet a commercial process, cellulase enzymes play a major role in the production of fuel alcohol from cellulose.

Ethanol from cellulose represents an enormous opportunity to make a transportation fuel that is an alternative to gasoline. Development of such a fuel is motivated by (i) an increased cleanliness of automobile exhaust, with decreased levels of carbon monoxide and nitrous oxides, (ii) a need for a fuel that does not contribute to an increase in the Greenhouse effect, (iii) the desire to decrease the dependence of the United States on imported petroleum, and (iv) the possibility of creating wealth in regions where cellulose is a prevalent natural resource.

Cellulose is converted to ethanol by making glucose and then fermenting the glucose to ethanol using yeast. Cellulose conversion has been studied for many years and has been carried out using (i) concentrated acid hydrolysis, (ii) dilute acid hydrolysis, or (iii) a combination of acid prehydrolysis and enzymatic hydrolysis. All three processes are under pilot-scale development. This discussion focuses on the acid prehydrolysis/enzymatic hydrolysis process.

For a cellulosic material such as wood chips or crop waste, cellulase enzymes cannot penetrate the structure and make glucose. A pretreatment is required to destroy the fiber structure and allow cellulase access to the substrate. Pretreatment is typically carried out at 180 – 250 ℃ for a few seconds to a few minutes in 0.5% to 2% sulfuric acid. The resulting material is of a muddy texture. An aqueous slurry of pretreated cellulose is made at 5% to 15% solids. Cellulase enzymes are added at a concentration of 5 to 25 filter paper units per gram of cellulose. The slurry is stirred, and the enzymatic hydrolysis is carried out for 4 to 7 days. At this point, most of the cellulose is converted to glucose, and the unhydrolyzed residue consists primarily of lignin.

The glucose syrup is removed from the residue by filtration and washing. The glucose is then fermented to ethanol, which is recovered by distillation.

The enzymatic hydrolysis process is carried out using crude Trichoderma cellulase at its optimum temperature and pH (50 ℃, pH 5). Although the crude Trichoderma cellulase is highly efficient at hydrolyzing cellulose, at present it is not efficient enough for a commercial ethanol process to be viable. One problem is the shortage of beta-glucosidase, which causes cellobiose, a very potent inhibitor of cellulase, to accumulate. Beta-glucosidase is not only in short supply, but it is highly inhibited by glucose.

In an alternative process designed to address the accumulation of cellobiose, the enzymatic hydrolysis and ethanol fermentation are carried out simultaneously, using the so-called SSF (simultaneous saccharification and fermentation) process. SSF is designed to remove the glucose by yeast fermentation, thereby decreasing end-product inhibition of the beta-glucosidase by glucose, and allowing the beta-glucosidase to continue to hydrolyze the cellobiose. The disadvantage of SSF is that it must be carried out at a temperature that is compatible with the yeast (optimum 28 ℃) and the cellulase (optimum 50 ℃). The compromise temperature usually used, 37 ℃, is suboptimal for both yeast and cellulase.

In addition to ethanol from cellulose, cellulase enzymes play a minor role in the production of ethanol from corn. In this process, most of the glucose is from starch. Cellulase enzymes offer the opportunity to increase the glucose yield by hydrolyzing a portion of the cellulose to glucose, as well as decreasing the viscosity of the ground corn.

9 Substrate Specificity and Synergy of Cellulase

The activity of the Trichoderma cellulase enzymes CBHI, CBHII, EGI, and EGII against several substrates is summarized in Tab. 2.2.

Tab. 2.2 Activity of cellulase enzymes

Substrate	Cellulase component[a]			
	CBHI	CBHII	EGI	EGII
Beta-glucan	−	+ + + +	+ + + + +	+ + +
Hydroxyethyl cellulose	−	+	+ + + +	+ +
Carboxymethyl cellulose	+	+ +	+ + + +	+ + + + +
Methylumbelliferyl cellotrioside	−	−	−	+
Para-nitrophenyl glucoside	−	−	−	+
Cellobiose	−	−	−	+
RBB xylan	−	−	+ + +	−
Crystalline cellulose	+ + + +	+ + +	+	+
Amorphous cellulose	+	+ + +	+ + + + +	+ + +

a: Relative Activity; None: −; Maximum: + + + + +.

Endoglucanases are the most active against soluble oligomers of glucose, such as beta-glucan or chemically-substituted cellulose, as well as amorphous cellulose. Cellobiohydrolases are most active against crystalline cellulose, with CBHI having little activity against other substrates. EGI is the least specific of the enzymes, as it has significant xylanase activity. None of the cellulase enzymes are especially active in solubilizing or hydrolyzing cellobiose.

There is a significant level of synergy among cellulase enzymes, resulting from their different, but complementary, modes of action. Among the four major Trichoderma cellulase components, every pair is synergistic except EGI and EGII. The synergy among the enzymes increases the degree of hydrolysis by more than two-fold over that achieved with individual enzymes.

10 Cellulase Assays

Although there are no assays of cellulase activity that are used universally, the most common assays are the Filter Paper(FPA) and carboxymethyl cellulose(CMCase) assays. The filter paper assay indicates the ability of the enzyme to produce reducing sugars from cellulose filter paper. This assay is particularly appropriate for crude Trichoderma cellulase or other cellulases with high levels of CBH components. It is not as appropriate for endoglucanases. The CMCase assay measures the ability of the enzyme to produce reducing sugars from soluble CMC. This assay is particularly useful in measuring the activity of many endoglucanases, which have especially high activity against this substrate. A variant of the CMCase assay is to measure the drop in viscosity of a CMC solution caused by enzyme action. This is a particularly sensitive measure of the initial attack of CMC by an enzyme.

Regardless of the assay used, the non-linearity of cellulase kinetics requires that the enzyme activity be measured based on a fixed level of conversion.

11 Cellulose Hydrolysis

When faced with characterizing the kinetic behavior of an enzyme or a complex of enzymes, one usually pulls out a textbook on Michaelis-Menten kinetics and applies it to the system at hand. For beta-glucosidase, which hydrolyzes the soluble substrate cellobiose to glucose, this approach is fine. Unfortunately, for cellulase enzymes producing cellobiose from cellulose, this exercise is inadequate.

Unit 2

The fact that cellulase enzymes act on an insoluble substrate, cellulose, moves the kinetics outside Michaelis-Menten on several counts. First of all, the enzyme can be adsorbed to the substrate or unadsorbed, but only the adsorbed enzyme acts on the cellulose. Even more puzzling is the substrate concentration. Do we count the entire substrate, or just that in close contact with the enzyme? Clearly, we have to start from first principles in characterizing the cellulase/cellulose system.

Unit 3

··· Part A ···

History of Biotechnology in Austria

1 Introduction

Biotechnology, if it can be considered a trade, can be traced back many centuries, when **wine** making, **brewing**, production of **vinegar** and **distilling** were important human skills. The history of biotechnology as an industry apparently begins in the early 19th century, parallel to the general change in **industrialization** in Europe and America.

Austria, i. e. the country now represented by the Republic of Austria, has contributed considerably to the development and progress of biotechnology. The beginning of this remarkable history may be traced back to the first decades of the 19th century although in this country earlier flourishing trades, such as wine making, brewing, distilling and the production of vinegar, were also practiced for many centuries.

In 1815, the **Vienna Polytechnic** Institute (Fig. 3. 1), now the Vienna University of Technology, was founded. From the very beginning biotechnological subjects were taught. The founder and first director of the Vienna Polytechnic Institute, Johann Josef Ritter von Prechtl (1778—1854), was the author of a renowned textbook of chemistry with special reference to chemical technology (1813) and, together with Altmütter and Karmarsch, was the editor of a 24-volume "Technological **Encyclopedia** or **Alphabetical** Handbook of Technology, Technical Chemistry and Mechanical Engineering" (1830). Teaching and research at this institute contributed considerably to the progress of Austrian industry at this time.

Unit 3

Fig. 3.1 The Vienna Polytechnic Institute near St. Charles Church

New Words

Austria	奥地利(欧洲中部国家)
biotechnology	生物工艺学
wine	葡萄酒,酒
brewing	酿造
vinegar	醋
distilling	蒸馏(作用)
industrialization	工业化,产业化
Vienna	维也纳(奥地利首都)
polytechnic institute	(多科性)工学院
von(= from of)	常加在姓前
encyclopedia	百科全书
alphabetical	依字母顺序的,字母的

2 The Vienna Process for Producing Baker's Yeast

An early example of Austria's historical role in biotechnology was the development of this process to produce baker's **yeast**. Until the 19th century, bakers obtained **dough**-leavening yeast mainly from local **breweries** which produced beer by the so-called top **fermentation**, where the yeast was recovered by **skimming** off the foam and separating the yeast mass by settling and **sieving**. When brewers changed to the more efficient bottom or lager fermentation, the resulting **bottom yeast** was inferior in quality and in quantity of supply. For example, in Vienna, the capital of the Austrian Empire, more than two hundred bakers seriously complained about this shortage. Distillers, although producing alcohol by a similar process using **top yeast**, were unable to suffice the increasing demand. Therefore, in 1847, the Federation of Industry of Lower Austria decided to offer a reward of 1,000 **gulden** together with a medal worth 50 **ducats** to the person who could produce an amount of 22.4 kg of yeast plus 40.74 L of alcohol from 193.8 kg of grain (values calculated from measures of that time). A further condition was that the competitor must prove his ability to supply and sell an amount of at least 5,000 kg of this yeast during a period of one year at normal market price.

The competition was won by Julius Reininghaus, a German chemist who had learned the **Dutch** art of yeast manufacture in Hannover and had offered his services to Adolf Ignaz Mautner, the owner of a brewing and distilling establishment in Vienna. Reininghaus was able to obtain yields even exceeding the requirements of the competition. Furthermore, he successfully introduced **maize** as a raw material for yeast production. He became Mautner's partner—and his brother Johann Peter became Mautner's son-in-law! Several additional production companies were founded and at the present time these two family names still represent renowned Austrian establishments. It was only about 70 years later that the Vienna Process was replaced by the more modern procedures involving **aeration** and feeding of the **carbon sources**.

New Words

yeast	酵母,发酵粉
dough	生面团

brewery	酿酒厂
fermentation	发酵
skim off	撇取(去);提出精华
sieve	筛,滤
bottom yeast	管底酵母
top yeast	表层酵母(菌)
gulden	基尔德(荷兰货币单位)
ducat	从前流通于欧洲各国的钱币达克特,硬币
Dutch	荷兰人,荷兰语;荷兰的
maize	玉米
aeration	通风
carbon source	碳源

❸ Technical Mycology, a Novel Field

Winemaking, brewing, distilling and the production of vinegar were already being taught at the Vienna Polytechnic Institute in the schedule of the school of special technical chemistry in 1816. Beginning with the work of Louis **Pasteur**, who established the scientific essence of these trades by studying and proving the biological and biochemical nature of fermentations, these fields developed into large industries with enormous production figures. Following the foundation of various research institutes, such as the Institute Pasteur in Paris, the Institute of Fermentation Research in **Copenhagen** and in Berlin, Austria also decided to establish a special university institute. This institute was founded at the Vienna Technical Institute in 1897 and still exists as the Institute of Biochemical Technology and **Microbiology** at the Vienna University of Technology. Its first director and professor was Franz Lafar (1865—1943) from Vienna (Fig. 3.2).

Lafar had worked at the **Agricultural Institute** of Hohenheim and as a lecturer at the Stuttgart **Technical Institute**. He had gained considerable reputation as the

Fig. 3.2 Franz Lafar(1865—1943), the founder of Technical Mycology

author of the two-volume "Handbook of Technical Mycology" in 1896 (English translation, 1898; Russian translation, 1903). This was followed by a five-volume second edition (1904 – 1914) which became a **standard source** of a novel discipline, Technical Mycology, a designation that he himself coined. Soon after, Technical Mycology was also taught at the Graz Technical Institute.

New Words

mycology	真菌学
Pasteur	巴斯德(Louis,1822—1895,法国化学家、细菌学家)
Copenhagen	哥本哈根
microbiology	微生物学
agricultural institute	农学院
technical institute	工业学院;初级技术学院;技术研究所;理工学院
standard source	标准源

4 Improvements in Distillery Practice

Besides his fame as one of the pioneers of the new field, Lafar also earned acclaim for the improvements he made in distillery practice. Distillers originally produced **alcohol** by purely **empirical** methods, using grain or potatoes as raw materials and the natural yeast **flora** within the distillery. Later, yeast was collected from the first batches of a production and used to seed successive batches, and this was carried out throughout the production campaign. Accordingly, severe **contaminations** were encountered. Through the work of the Berlin Institute (Delbrueck), pure culture yeast ("Kunsthefe") became available and it was especially recommended that this "artificial" yeast be **propagated** under conditions of "natural **pure culture**", i.e. adapted to the conditions of the substrates being processed in the respective distilleries.

In order to counteract contamination, mainly from **butyric acid** bacteria, it was common practice to maintain a **spontaneous lactic acid** fermentation, which was introduced by the natural bacterial flora of the machine and the environment, and it was hoped that this would remain active throughout the season. In 1893, in an attempt to create **optimum conditions** for this protective fermentation, Lafar isolated the most potent bacterial strain from an actively souring yeast seed culture and introduced this culture successfully to all the distilleries in the Hohenheim area during the following campaigns. In 1896, after this method had been adopted in the whole Württemberg area, he published his findings designating the organism as **Bacillus** acidificans longissimus, but only mentioned to provide a more accurate description. At the same time, and in the same journal following Lafar's paper, Leichmann described the isolation of a similar strain, which he designated Bacillus delbruecki, and this was the name to subsist for the apparently identical strain. The designation Bac. Acidificans(Bac. delbruecki) was used by distillers for some time, but nowadays the literature only mentions **Lactobacillus** delbrueckii, in particular, as the organism of the current industrial lactic acid fermentation process.

New Words

alcohol	酒精,酒
empirical	完全根据经验的,实验式
flora	细菌群落
contamination	污染,污染物
propagate	繁殖,传播,宣传
pure culture	纯培养物,纯粹培养
butyric acid	丁酸
spontaneous	自发的,自然产生的
lactic acid	乳酸
optimum condition	最佳工况(条件);最适条件
bacillus	杆状菌,细菌
lactobacillus	乳酸菌

  5 The Advent of Plant Cell Culture

Since **plant tissue** culture has become a potential biotechnological field, it is justified to investigate the past of this valuable tool. As early as 1839, Schwann suggested that plant cells should be considered **totipotent**. This means that each living cell of plant tissue is able to develop into a whole organism provided the cell is maintained in a proper environment, esp. with respect to nutrition.

The first experiments with **fragmented** plant tissues resulting in the formation of actively **multiplying** cells were performed before the turn of the 20th century. The Austrian scientist Rechinger(1893)even tried to determine and to define the "limits of divisibility" of various plant materials. It was the great Austrian biologist Gottlieb Haberlandt, however, who in 1902 established the foundations of plant tissue culture. Unlike Rechinger, Haberlandt believed that it was even possible to propagate isolated plant cells. Although his experiments were of limited success, his merit as the founder of this discipline has been fully acknowledged during this century(see, e. g. Krikorian and Bequam, 1969) and quite recently, in 1998, this fact was celebrated in an international **symposium**.

By choosing more suitable plant material, **root tips**, and better nutrient media, excellent results were achieved-first by Gautheret in 1934. Since then, plant cell culture has become a fruitful discipline within biotechnology, with **manifold** economic potential. This includes the production of various products of **secondary metabolism** as well as e. g. **transgenic** crops.

Obviously, the **photosynthetic** potential of plants with respect to the production of biomass as a renewable resource in sustainable production cycles has found actual attention and has been defined in many recent national and international research programs. A special variant of such endeavors has been formulated as "New Phytotechnology" by the Austrian group of Othmar Ruthner and coworkers and this will be dealt with in the following section.

New Words

plant tissue	植物(性)组织
totipotent	(细胞)全能的
fragmented	成碎片的,片断的
multiply	繁殖
symposium	讨论会,座谈会
root tip	根尖
manifold	多方面的
secondary metabolism	次级代谢
transgenic	基因改造的,基因被改变的(gene-altered)
photosynthetic	光合的;促进光合作用的
phyto-	表示"植物"

Unit 3

6 New Phytotechnology

The basic idea may be defined as attempts to utilize light (solar) energy in a **controlled artificial environment** by establishing some kind of plant factory enabling **continuous production** of any kind of plant independent of site and season. This may be realized on a large (industrial) scale by a three-dimensional driven conveyor system in a closed environment illuminated by a fixed light-**lattice**. The **environmental conditions** in such systems (Fig. 3.3) may be optimized according to the specific requirements of the crop to be produced. Continuous industrial plant production may serve not only to provide fresh vegetables, green **fodder**, and various plant material for pharmaceutical purposes (e. g. **Digitalis lanata**), but also for the propagation of seedlings or shoots for mass cultivation, e. g. for short rotation forestry to produce **renewable** energy resources.

It has been claimed by the producers of these systems (Ruthner Pflanzentechnik Ltd. and Maschinenfabrik Andritz Ltd.) that, for example, the water requirements in such facilities are only 2% of that in conventional European fodder production. Fertilizer requirements are much lower than in conventional economies and the **pesticide** demand is reduced considerably. This would suggest its application not only in **arid zones** but also in space.

Fig. 3.3　Continuous industrial plant production system (O. Ruthner)

It should be noted at this point that historically the idea of systematically investigating plants as sources of various raw materials goes back to the great Austrian scientist Julius von Wiesner (1838 – 1916), who established the science of natural materials (Rohstofflehre) with his famous book, "Die Rohstoffe des Pflanzenreiches", in 1873. Haberlandt was one of his students.

New Words

controlled environment	受(可)控环境
continuous production	连续生产,流水生产
lattice	格子,格子格构,格构制品
environmental condition	环境条件[状况]
fodder	饲料,草料
digitalis	洋地黄
Digitalis lanata	柔毛洋地黄
renewable	可更新的,可恢复的
pesticide	杀虫剂
arid zone	干旱带

7 An Important Role in Citric Acid Fermentation

Commercial citric acid fermentation began with the pioneering work of Currie(1917) in the United States, who initiated the first successful industrial production of **citric acid** in 1923 with Chas. Pfizer in **Brooklyn**. This venture almost demolished the **market position** of citric acid from **citrus fruits** held by Italy. Soon after, attempts were made to establish respective plants in Europe. Interestingly, the first **patent** was applied for in Austria in 1923 by J. Szücs from Vienna and granted in 1925. Szücs offered his knowledge to a company in **Prague**. As early as 1928, a plant was built at Kaznéjow near Plzen, and this plant went into production using for the first time **molasses** as raw material, according to Szücs's patents. It was in this plant that the treatment of molasses with hexacyanoferrate was invented, a method still in use in industries using less pure raw materials, and which has been studied intensively for decades by several research groups. Today, Austria is one of the most prominent producers of citric acid in the world.

Unit 3

New Words

citric acid	柠檬酸
Brooklyn	布鲁克林(美国纽约市西南部的一区)
market position	市场地位
citrus fruit	柑橘类的水果
patent	专利权,专利品;专利的;取得……的专利权,授予专利
Prague	布拉格
molasses	糖蜜

 Further Improvements in Yeast Production

About one hundred years after the invention of the **Viennese** process for baker's yeast production, several improvements to this art were again made in Vienna. W. Vogelbusch, a process engineer and owner of a consulting firm working with Hefefabriken Mautner Markhof, invented several **rotating** aeration devices to replace the conventional static aerators in baker's yeast production. It had been known since the basic investigations of **Pasteur** that **oxygen** suppresses fermentation (Pasteur effect), and this had given rise to the so-called "Zulauf" processes as a new technology of yeast manufacture, comprising low feed rates of the carbon source together with high aeration rates.

The new rotating aerators of **Vogelbusch**, especially the so-called "**dispergator**" (Fig. 3.4 (a), (b)) provided higher oxygen transfer rates, thus saving air and enabling higher feed rates of the **carbon sources** resulting in higher productivities. These feed rates, in turn, were usually adjusted according to empirical schedules owing to the **logarithmic** law of yeast growth. An attempt was made to keep the concentration of the carbon source as low as possible to avoid excessive **aerobic fermentation** producing alcohol which would get lost via the exhaust air.

This was the starting point for a further improvement in the regulation of the carbon source feed rate. By measuring the ethanol content of the exhaust air (representing the ethanol concentration in the mash according to Henry's law), using **catalytic oxidation** of the ethanol and converting the heat generation into an electrical signal, the feed rate could be adjusted elegantly to the oxygen demand, i.e. the oxygen transfer property of the aerator. The so-called "Autoxymax" principle of Vereinigte Hefefabriken Mautner Markhof is in use in many yeast plants all over the world. The initial exhaust gas sensor has now been replaced by a system derived from common smoke detection devices.

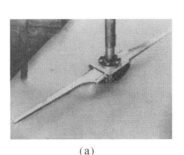 

(a)   (b)

Fig. 3.4 (a) **Vogelbusch dispergator**(courtesy of Aktiengesellschaft Kühnle, Kopp and Kausch, Frankenthal, Germany); (b) **Vogelbusch dispergator with cooling device and baffles**(courtesy of Vogelbusch GmbH, Vienna)

Yet another improvement was of great influence on the economics of yeast production: The separation of the yeast from the spent mash was performed by centrifugation and subsequent dehydration of the resulting yeast cream in a frame press. Only the application of frame presses allowed dry substance values of about 27% to be attained, this being the desired standard with respect to handling properties and shelf-life. Attempts to replace frame presses with rotating drum filters showed that such dry substance values were barely achievable. The problem was solved in an ingenious way by K. V. Rokitansky and E. Küstler.

Rokitansky, one of the chief chemists in the above-mentioned establishment, had studied not only chemistry but also **botany** with the famous **botanist** F. Weber at Graz University. As many readers know, one of the favorite objects of **introductory** microscopic **courses** is the **onion** cell (Allium cepa), where in particular the phenomena of cell **turgor** and cytorrhysis can be studied. When, years after this, Rokitansky was reasoning about the negative results with a **rotating** drum **filter** to separate yeast suspensions, he remembered his observations with cytorrhysis experiments, demonstrating the dehydrating action of e. g. salt gradients on cells. Together with Küstler, he developed a method of dehydrating yeast creams on a rotating drum filter by pretreating the yeast cream with a **sodium chloride** solution and subsequently separating the dehydrated yeast cells on the filter. Adhering salt solution could be removed by quickly **spraying** with water in a subsequent zone of the filter thus avoiding **rehydration** of the cells. With this invention, dry substance values exceeding 30% could be achieved, which facilitated subsequent adjustment of particular dry substance values and enabled yeast to be provided with improved shelf-life.

Together with a process of combined yeast and ethanol production, the so-called KOMAX process, in which the propagation of yeast is performed in a way that a definable amount of yeast from the ethanol producing stage can be used as seed-yeast for the successive baker's yeast stage, the inventions mentioned above constitute most of the advanced technology of yeast manufacture today which, at least in part, is applied in many countries.

New Words

Viennese	维也纳人;维也纳人的,维也纳文化的
rotating	旋转式喷灌器
Pasteur	巴斯德
oxygen	氧
Vogelbusch	奥地利奥高布殊公司
dispergator	解胶剂
carbon source	碳源
logarithmic	对数的
aerobic fermentation	需氧(气)发酵
catalytic oxidation	催化氧化
centrifugation	离心法;离心过滤
dehydration	脱水
shelf-life	(包装食品的)货架期,保存限期
botany	植物学
botanist	植物学家
introductory course	预备课程;导论课;入门课
onion	洋葱
turgor	细胞(组织)的膨胀

rotating filter 滤色转盘
sodium chloride 氯化钠
spraying 喷雾
rehydration 再水化,再水合

9 Ergot Alkaloids

Brief mention should be made of Austria's part in the history of producing these substances. Through the centuries, ergot **alkaloids** were the causative agents of severe **epidemic diseases**, **ergotism**. Typical **manifestations** were **convulsive** and **gangrenous** ergotism, and these were handed down under various names due to their striking actions, e. g. ignis sacer(holy fire) or plaga ignis or pestilens ille **morbus**, etc. It appears that the beneficial actions of ergot alkaloids, namely to enhance **muscle contractions**, esp. to provoke uterus contractions during childbirth, were utilized even before the details of ergotism were known.

Ergot alkaloids are formed by all known(about 50) species of the **fungus Claviceps** and, to a lesser extent, also by some other fungi, e. g. **Aspergillus** and **Penicillium.** Claviceps infects mainly grasses, of which **rye** and other cereals appear as typical examples being responsible for the former epidemic outbreaks of ergotism mentioned above. For medical uses the **sclerotia** of the fungus were collected from these **cereals**, especially in rye fields, and processed in small pharmaceutical establishments. The first **clinically** used compound, ergotamin, was discovered by Stoll in 1918. Obviously, there was increasing interest in developing more productive and controllable methods of production, especially since it became apparent that yield as well as type of alkaloid or alkaloid group was rather strain-specific and dependent on environmental conditions.

This was the beginning of the so-called **parasitic** production of ergot alkaloids, which was developed in Hungary and improved in Austria and Switzerland. The essence of these methods was to **inoculate** ears of rye before or at the time of flowering with a conidia suspension of Claviceps by an injection device causing small lesions, e. g. using inverted sewing needles with the ears of the needles as a suitable reservoir for the necessary amount of suspended conidia for infection. Yields per acre of ergot alkaloids could be increased considerably and uniform alkaloid moieties could be obtained.

Today, this method has been replaced by fermentation processes, enabling the production of a wide spectrum of specific compounds by the most suitable strains under the most precise production schedules.

Unit 3

New Words

ergot	麦角,麦角碱;麦角菌
alkaloid	生物碱,植物碱基
epidemic disease	流行病
ergotism	麦角中毒
manifestation	显示,表现
convulsive	起痉挛的,痉挛性的
gangrenous	坏疽的,腐败的,脱疽的
morbus	(＝disease)(疾)病
muscle contraction	肌肉收缩
fungus	菌类,蘑菇
Claviceps	麦角菌
aspergillus	曲霉菌
penicillium	青霉菌
rye	裸麦,黑麦
sclerotia	菌核
cereal	谷类食品,谷类
clinical	临床的,病房用的
parasitic	寄生的
inoculate	接种,菌体培养

10 The Submerged Vinegar Process

Shortly after the Second World War, in a period of many changes in the economic situation in Austria, two chemists met by chance in an office in Upper Austria, one of which, Heinrich Ebner, was working in a vinegar plant, whereas the other, Otto Hromatka, an organic chemist with a strong pharmaceutical background, was in search of a new field of activity. Reasoning about the fact that **vinegar** was not produced by a submerged process, the two scientists decided to try to transfer the old-fashioned trickling process into a modern **submerged fermentation** technology.

The essence of the trickling process (generator process) is to charge a reactor, filled with e. g.

wood shavings with an adhering active population of acetic acid bacteria, from the top with wine or beer or **diluted** ethanol containing a certain amount of vinegar(in order to avoid **overoxidation**) while aerating from the bottom. In the old Schuezenbach process, vinegar was produced in one step and withdrawn at the bottom. In the more modern generator process with higher reactor volumes causing internal **overheating**, the necessity of cooling required shorter residence times. This was accomplished by **circulating** the mash and cooling it outside the reactor.

Hromatka and Ebner observed that active **acetic acid bacteria** in a submerged system were extremely sensitive to interruptions in the aeration. They found that an actively oxidizing bacterial population could not be obtained by the usual procedures of **inoculating with** a normal bacterial pure culture, e. g. from an **agar medium**. This could be achieved, however, by placing wood shavings from a working generator into a continuously aerated mash until a certain number of cells became **suspended** in the mash and began to multiply. In this way, the submerged vinegar process was developed. Subsequently, the know-how was merged with that of Frings Ltd., Bonn, the company now producing this type of vinegar plant.

Present reactors, so-called acetators (Fig. 3.5), are equipped with self-priming aerators (guarded by an emergency power station), an efficient cooling system and analyzers to determine the composition of the mash in situ. The advantages, as compared with the preceding generators,

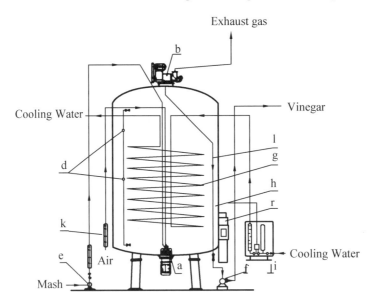

Fig. 3.5 **Modern acetator for the production of vinegar(courtesy of Frings, Bonn)**

are much higher productivities, the possibility of producing purer vinegar (acetic acid), e. g. from pure ethanol, and no transient batches when changing the raw material. The only disadvantage is the fact that **clarification** of the resulting vinegar is more expensive. A great number of plants all over the world have changed to this efficient process.

New Words

vinegar	醋
submerged fermentation	液面下发酵,底部发酵,深层发酵
acetic acid	乙酸,醋酸
diluted	无力的;冲淡的
overoxidation	过氧化
overheating	过热,热度过高
circulating	循环,流通
acetic acid bacteria	醋酸菌
inoculate with	灌输
agar medium	琼脂培养基
suspended	悬浮的
clarification	澄清,净化

 The Penicillin V Story

The discovery of **penicillin** by Alexander Fleming and its large-scale production, realized by the famous **Oxford** group of scientists and a consortium of US companies during World War II, has changed our **life expectancy** almost unbelievably. No wonder that the story of this great discovery has been told many times. In contrast, the story of the discovery of the first acid-stable, oral penicillin is less well known—in some of the various sources it is even neglected.

One of the few disadvantages of the common penicillin, designated penicillin G, was its **lability** under acidic conditions. Therefore, penicillin G could not be administered orally. Moreover, it was difficult to build up stable blood levels because, parallel to its low **toxicity**, penicillin was excreted within a few hours after injection inevitably demanding frequent treatment. It was therefore acclaimed as a considerable achievement when the desired oral penicillin was discovered.

Soon after the end of World War II, a small plant was established in the Tyrol, then part of the French occupied zone of Austria, in a closed-down **brewery** of the Austrian Brewing Corporation: the Biochemie Kundl GmbH. Research in this establishment was entrusted to Richard Brunner (later professor at the Vienna University of Technology), who had witnessed the first experiments of a German research group under K. Bernhauer in Prague to produce penicillins during World War II.

Due to the special situation in the post-war era, the implementation of this endeavor was extremely difficult. With the help of a French chemist, Captain Rambaud, of the French occupation forces, a small team of scientists and engineers succeeded in producing sufficiently pure penicillin within a rather short period of time (1948). Problems of equipment were solved by using various redundant military materials, e. g. V2 missile containers as liquid vessels, self-produced fermenters stirred with the help of motors of submarines and aerated by compressors powered by motors of German Tiger tanks. The necessary pipes were obtained from a bombed Innsbruck café. Since corn-steep liquor was not available, **yeast extract** had to be used, and **whey** had to serve as a substitute for lactose. Even the necessary butanol for the preparation of the extractant had to be produced by installing a **butanol** fermentation.

Obviously, one of the major obstacles was the frequent occurrence of microbial contaminations during fermentations which destroyed many valuable batches. In the endeavor to counteract such contaminations, Ernst Brandl (Fig. 3.6), working on his dissertation in the microbiological and fermentation laboratory, tried to add 2-phenoxyethanol, a compound mainly in use as a **preservative** in **cosmetic** preparations, to the fermentation medium. The surprising effect was a significant discrepancy between the results of **bioassays** and those of chemical (**iodometric**) determinations in the resultant fermentation broth. This **phenomenon** was studied by Hans Margreiter (Fig. 3.6), working with Brunner in the chemical research laboratory of the plant. Surprisingly, when trying to isolate the penicillin moiety by extraction with **diisopropyl ether**, he observed that **crystalline** precipitates with penicillin activity had been formed in the acid aqueous phase after prolonged standing. Soon it was realized that a novel, acid-stable penicillin had been discovered, apparently due to the ability of the fungus to utilize the added **phenoxyethanol** as side-chain precursor after its oxidation to phenoxyacetic acid. The respective patent application was filed in 1952.

The possibility of **oral administration** of this novel penicillin, designated as penicillin V (phenoxymethyl penicillin), together with its low toxicity, paved the way to high **dosage therapy**, which was successfully introduced by K. H. Spitzy in Vienna.

Unit 3

Fig. 3.6 E. Brandl and H. Margreiter(courtesy of Biochemie Kundl GmbH, Kundl, Tyrol)

Considerable disappointment arose when it was found that the formation of phenoxy methyl penicillin had already been described, and patented, by O. Behrens from E. Lilly in the US, who had, however, not recognized its acid stability. Successive negotiations were able to settle this problem in a most satisfactory way for both sides and both companies were able to acquire leading positions as producers of oral penicillin. Biochemie Kundl GmbH., now part of Novartis Ltd., is still one of Austria's biotechnical companies most renowned for its production of antibiotics, enzymes and other specialties such as e. g. **cyclosporin**. Quite recently (1998), the company acquired the fermentation facilities of Hoechst Marion Roussel at Frankfurt to enlarge its production capacity about twofold.

About one third of the world production of oral penicillin comes from Austria.

New Words

penicillin	青霉素(盘尼西林)
Oxford	牛津(英国城市),牛津大学
life expectancy	平均寿命(= expectation of life)
lability	不稳定(性),易变(性)

toxicity	毒性
brewery	酿酒厂
liquor	液体,汁,酒精饮料,(药)溶液
yeast extract	酵母抽提物,酵母膏
whey	乳清
butanol	丁醇
preservative	防腐剂
cosmetic	化妆品;化妆用的
bioassay	生物测定,生物鉴定
iodometric	碘量法的
phenomenon	现象
diisopropyl ether	二异丙醚
crystalline	晶体的,晶体物的
phenoxy	含苯氧基的
ethanol	乙醇,酒精
oral administration	口服
dosage	剂量,配药,用量
therapy	治疗
cyclosporin	环胞霉素

Unit 3

··· Part B ···

Genetic Engineering

 Introduction

Genetic engineering, also called genetic modification, is the direct manipulation of an organism's genome using biotechnology. It is a set of technologies used to change the genetic makeup of cells, including the transfer of genes within and across species boundaries to produce improved or novel organisms. New DNA may be inserted in the host genome by first isolating and copying the genetic material of interest using molecular cloning methods to generate a DNA sequence, or by synthesizing the DNA, and then inserting this construct into the host organism. Genes may be removed, or "knocked out", using a nuclease. Gene targeting is a different technique that uses homologous recombination to change an endogenous gene, and can be used to delete a gene, remove exons, add a gene, or introduce point mutations.

The polymerase chain reaction (PCR) is a laboratory technique that can produce many copies of a gene or segments of a gene, which makes studying the gene much easier. A specific segment of deoxyribonucleic acid (DNA), such as a specific gene, can be copied (amplified) in a laboratory. Starting with one DNA molecule, at the end of 30 doublings (only a few hours later) about a billion copies are produced.

Various methods may be used to find (probe) changes in genes. A gene probe can be used to locate a specific part of a gene (a segment of the gene's DNA) or a whole gene in a particular chromosome. Probes can be used to find normal or mutated segments of DNA. A DNA segment that has been cloned or copied becomes a labeled probe when a radioactive atom or fluorescent dye is added to it. The probe will seek out its mirror-image segment of DNA and bind to it. The labeled probe can then be detected by sophisticated microscopic and photographic techniques. With gene probes, a number of disorders can be diagnosed before and after birth. In the future, gene probes will probably be used to test people for many major genetic disorders simultaneously.

An oligonucleotide is a chain of bases (nucleotides). Sometimes these chains are missing or have duplicate segments of DNA. An oligonucleotide array is used to identify deleted or duplicated segments of DNA in specific chromosomes. In anoligonucleotide array, DNA from a person is compared to a reference genotype using many oligonucleotide probes. Like some gene probes, a fluorescent dye is added to the oligonucleotide probes. If a segment is missing, the probes detect a decreased amount of the fluorescent dye. If a segment is duplicated or tripled, the probes detect an increased amount of the fluorescent dye. These probes can be used to test the entire genotype.

An organism that is generated through genetic engineering is considered to be a genetically modified organism (GMO). The first GMOs were bacteria generated in 1973 and GM mice in 1974. Insulin-producing bacteria were commercialized in 1982 and genetically modified food has been sold since 1994. GloFish, the first GMO designed as a pet, was first sold in the United States in December 2003.

Genetic engineering techniques have been applied in numerous fields including research, agriculture, industrial biotechnology, and medicine. Enzymes used in laundry detergent and medicines such as insulin and human growth hormone are now manufactured in GM cells, experimental GM cell lines and GM animals such as mice or zebrafish are being used for research purposes, and genetically modified crops have been commercialized.

2 Transcription Expresses Genes

Gene expression involves making an RNA copy of information present on the DNA, that is, transcribing the DNA. Making RNA involves uncoiling the DNA, melting the strands at the start of the gene, making an RNA molecule that is complementary in sequence to the template strand of the DNA with an enzyme called RNA polymerase, and stopping at the end of the gene. The newly made RNA is released from the DNA, which then returns to its supercoiled form.

An important issue in transcription is identifying the right gene. Which gene needs to be decoded to make protein? There are different types of genes. Some are housekeeping genes that encode proteins that are used all the time. Other genes are activated only under certain circumstances. For instance, in E. coli, genes that encode proteins involved with the utilization of lactose are expressed only when lactose is present. The same principle applies to the genes for using other nutrients. Various inducers and accessory proteins control whether or not these genes are expressed ormade into RNA and will be discussed in more detail in upcoming sections.

Unit 3

The final product encoded by a gene is often a protein but may be RNA. Genes that encode proteins are transcribed to give messenger RNA, which is then translated to give the protein. Other RNA molecules, such as tRNA, rRNA, and snRNA, are used directly (i.e., they are not translated to make proteins). Some RNA molecules, such as large-subunit rRNA, are called ribozymes and can catalyze enzymatic reactions. Though most of the time, genes ultimately code for a protein via an mRNA intermediate. The coding region of a gene is sometimes called a cistron or a structural gene and may encode a protein or a nontranslated RNA. In contrast, an open reading frame (ORF) is a stretch of DNA (or the corresponding RNA) that encodes a protein and therefore is not interrupted by any stop codons for protein translation (see later discussion).

The next issue is finding the start site of the gene. Every gene has a region upstream of the coding sequence called a promoter (Fig. 3.7). RNA polymerase recognizes this region and starts transcription here. Bacterial promoters have two major recognition sites: the −10 and −35 regions. The numbers refer to their approximate location upstream of the transcriptional start site. (By convention, positive numbers refer to nucleotides downstream of the transcription start site and negative numbers refer to those upstream.) The exact sequences at −10 and −35 vary, but the consensus sequences are TATAA and T TGACA, respectively. When a gene is transcribed all the time or constitutively, then the promoter sequence closely matches the consensus sequence. If the gene is expressed only under special conditions, activator proteins or transcription factors are needed to bind to the promoter region before RNA polymerase will recognize it. Such promoters rarely look like the consensus.

Just after the promoter region is the transcription start site. This is where RNA polymerase starts adding nucleotides. Between the transcription start site and the ORF is a region that is not made into protein called the 5′ untranslated region (5′ UTR). This region contains translation regulatory elements like the ribosome binding site. Next is the ORF, where no translational stop codons are found. Then there is another untranslated region after the ORF, known as the 3′ untranslated region (3′ UTR). Finally that comes the termination sequence where transcription stops.

Bacterial RNA polymerase is made of different protein subunits. The sigma subunit recognizes the -10 and -35 regions and the core enzyme catalyzes the RNA synthesis. RNA polymerase only synthesizes nucleotide additions in a 5′ to 3′ direction. The core enzyme has four protein subunits, a dimer of two α-proteins, a β-protein, and a related β′ subunit. The β and β′ subunits form the catalytic site, and the α subunit helps recognize the promoter. The 3D structure of RNA polymerase shows a deep groove that can hold the template DNA, and a minor groove to hold the growing RNA.

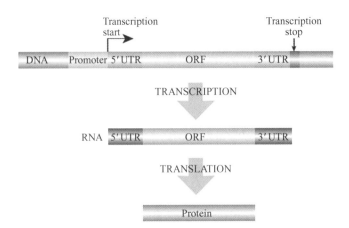

Fig. 3.7 The Structure of a Typical Gene

Genes are regions of DNA that are transcribed to give RNA. In most cases, the RNA is translated into protein, but some RNA is not. The gene has a promoter region plus transcriptional start and stop points that flank the actual message. After transcription, the RNA has a 5' untranslated region (5' UTR) and 3' untranslated region (3' UTR), which are not translated; only the ORF is translated into protein.

3 Replication of DNA

Replication copies the entire set of genomic DNA, so that the cell can divide in two. During replication, the entire genome must be uncoiled and copied exactly. This elegant process occurs extremely fast in E. coli, where DNA polymerase copies about 1,000 nucleotides per second. Although the process is slower in eukaryotes, DNA polymerase still copies 50 nucleotides per second. Many biotechnology applications use the principles and ideas behind replication.

In order to maintain the integrity of an organism, the entire genome must be replicated identically. Even for gene creatures such as plasmids, viruses, or transposons, replication is critical for their survival. The complementary two-stranded structure of DNA is the key to understanding how it is duplicated during cell division. The double-stranded helix unwinds, and the hydrogen bonds holding the bases together melt apart to form two single strands. This Y-shaped region of DNA is the replication fork (Fig. 3.8). Replication starts at a specific site called an origin of replication (ori) on the chromosome. The origin is called oriC on the E. coli chromosome and covers about 245 base pairs of DNA. The origin has mostly AT base pairs, which require less energy to break than GC pairs.

Unit 3

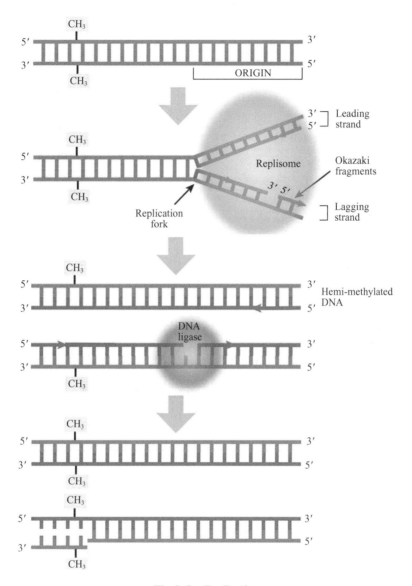

Fig. 3.8 Replication

Replication enzymes open the double-helix around the origin to make it single-stranded. DNA polymerase adds complementary nucleotides to each side in a 5′ to 3′ direction; therefore, one strand is synthesized continuously (leading strand) and the other strand (lagging strand) is synthesized in short pieces called Okazaki fragments. DNA ligase seals any nicks or breaks in the phosphate backbone. Finally, methylases add methyl groups to the newly synthesized strands.

Once the replication fork is established, a large assembly of enzymes and factors assembles to synthesize the complementary strands of DNA (see Fig. 3.8). DNA polymerase starts synthesizing the complementary strand on one side of the fork by adding complementary bases in a 5′ to 3′ direction. This strand is synthesized continuously because there is always a free 3′-OH group. This strand is called the leading strand. Because DNA polymerase synthesizes only in a 5′ to 3′ direction, the other strand, called the lagging strand, is synthesized as small fragments called Okazaki fragments. The lagging strand fragments are ligated together by an enzyme called DNA ligase. Ligase links the 3′ – OH and the 5′ – PO_4 of neighboring nucleotides, forming a phosphodiester bond. The final step is to add methyl (CH_3) groups along the new strand. The original double-stranded helix is now two identical double-stranded helices, each containing one strand from the original molecule and one new strand. This is why the process is called semiconservative replication.

4 Comparing Replication in Gene Creatures, Prokaryotes, and Eukaryotes

Although the basic mechanism for replication is the same for most organisms, the timing, direction, and sites for initiation and termination are variable. The major differences in replication occur mainly because of the special challenges posed by circular and linear genomes. Normal DNA replication occurs bidirectionally in prokaryotes and eukaryotes, whether the genome is linear or circular. Two replication forks travel in opposite directions, unwinding the DNA helix as they go. In bacteria such as *E. coli*, there is only one origin, oriC, and replication occurs in both directions around the circular chromosome until it meets at the other side, the terminus, terC. Halfway through this process, the chromosome looks like the Greek letter θ; therefore, this process is often called theta-replication (Fig. 3.9). The single circular chromosome then becomes two. Theta replication is also used by many plasmids, such as the F plasmid of E. coli, when growing and dividing asexually (as opposed to transferring its genome to another cell via conjugation).

Some plasmids and many viruses replicate their genomes by a process called rolling circle replication (Fig. 3.10). At the origin of replication, one strand of the DNA is nicked and unrolled. The intact strand thus rolls relative to its partner (hence "rolling circle"). DNA is synthesized from the origin using the circular strand as a template. As DNA polymerase circles the template strand, the new strand of DNA is base-paired to the circular template. Meanwhile the other parental strand is dangling free. This dangling strand is removed, ligated to form another circle, and finally a second strand is synthesized. This results in two rings of plasmid or viral DNA, each with one strand from the original molecule and one newly synthesized strand.

Unit 3

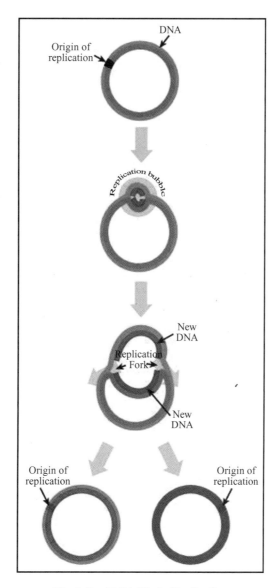

Fig. 3.9 (left) Theta Replication

In circular genomes or plasmids, replication enzymes recognize the origin of replication, unwind the DNA, and start synthesis of two new strands of DNA, one in each direction. The net result is a replication bubble that makes the chromosome or plasmid look similar to the Greek letter theta (θ). The two replication forks keep moving around the circle until they meet on the opposite side.

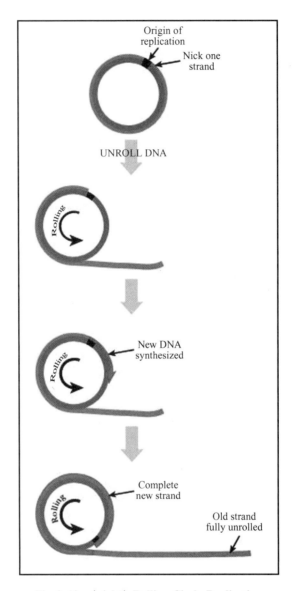

Fig. 3.10 (right) Rolling Circle Replication

During rolling circle replication, one strand of the plasmid or viral DNA is nicked, and the broken strand (pink) separates from the circular strand (purple). The gap left by the separation is filled in with new DNA starting at the origin of replication (green strand). The newly synthesized DNA keeps displacing the linear strand until the circular strand is completely replicated. The linear single-stranded piece is fully "unrolled" in the process.

Unit 3

Long linear DNA molecules such as human chromosomes pose several problems for replication. The ends pose a particularly difficult problem because the RNA primer is synthesized at the very end of the chromosome. When the RNA primer is removed by an exonuclease (MF1), there is no upstream 3′ – OH for addition of new nucleotides to fill the gap. (In eukaryotes, there is no equivalent to the dual-function DNA polymerase I. A separate exonuclease, MF1, removes the RNA primers, and DNA polymerase δ fills in the gaps.) Over successive rounds of replication, the ends of linear chromosomes get shorter and shorter. Special structures called telomeres are found at the tips of each linear chromosome and prevent chromosome shortening from affecting important genes. Telomeres have multiple tandem repeats of a short sequence (TTAGGG in humans). The enzyme telomerase can regenerate the telomere by using an RNA template to synthesize new repeats. This only happens in some cells; in others, the telomeres shorten every time the cell replicates its DNA. One theory regards telomere shortening as a molecular clock, aging the cell, and eventually triggering suicide.

The length of linear chromosomes also poses a problem. The time it takes to synthesize an entire human chromosome would be too long if replication began at only one origin. To solve this issue, there are multiple origins, each initiating new strands in both directions. These are elongated until they meet the new strands from the other direction.

5 PCR in Genetic Engineering

PCR allows the scientist to clone genes or segments of genes for identification and analysis. PCR also allows the scientist to manipulate a gene that has already been identified. Various modified PCR techniques allow scientists to hybridize two separate genes or genes segments into one, delete or invert regions of DNA, and alter single nucleotides to change the gene and its encoded protein in a more subtle way.

PCR can make cloning a foreign piece of DNA easier. Special PCR primers can generate new restriction enzyme sites at the ends of the target sequence (Fig. 3.11). The primer is synthesized so that its 5′ end has the desired restriction enzyme site, and the 3′ end has sequence complementary to the target. Obviously, the 5′ end of the sequence information is unknown, but rather than using degenerate primers, primers with a randomly chosen sequence are made. The length of a primer determines how often it will bind within the target DNA. If a particular primer were, say, 5 bases long, it would bind once on average every 4^5 bases $= 4 \times 4 \times 4 \times 4 \times 4 = 1,024$ bases. If the target DNA were a sample from a large genome, such a primer would bind far too many times. In practice longer primers of around 10 bases are often suitable.

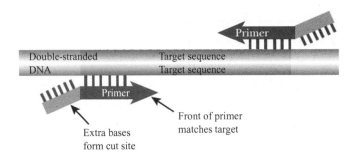

Fig. 3.11 Incorporation of Artificial Restriction Enzyme Sites
Primers for PCR can be designed to have nonhomologous regions at the 5' end that contain the recognition sequence for a particular restriction enzyme. After PCR, the amplified product has the restriction enzyme sites at both ends. If the PCR product is digested with the restriction enzyme, this generates sticky ends that are compatible with a chosen vector.

The random primer is mixed with nucleotides, Taq polymerase, and each of the target DNA samples as for normal PCR reactions. In order to amplify any target DNA fragments, two of the random primers must bind to the target DNA, on opposite strands, usually within a few thousand bases. The results of the two PCR reactions are compared using gel electrophoresis. The number and size of PCR products will vary for the two samples. If two organisms are very closely related, then their DNA will be close in sequence. Hence, the PCR products will be synthesized that has nucleotide mismatches in the middle region of the primer. The primer will anneal to the target site with the mismatch in the center. The primer needs to have enough matching nucleotides on both sides of the mismatch so that binding is stable during the PCR reaction. The mutagenic primer is paired with a normal primer. The PCR reaction then amplifies the target DNA incorporating the changes at the end with the mutagenic primer. These changes may be relatively subtle, but if the right nucleotides are changed, then a critical amino acid may be changed. One amino acid change can alter the entire function of a protein. Such an approach is often used to assess the importance of particular amino acids within a protein.

6 Getting Cloned Genes into Bacteria by Transformantion

Once the gene of interest is cloned into a vector, the construct can be put back into a bacterial cell through a process called transformation (Fig. 3.12). Here the "naked" DNA that was constructed in the laboratory is mixed with competent E. coli cells. To make the cells competent, that is, able to take up naked DNA, the cell wall must be temporarily opened up. In the laboratory, E. coli cells are treated with high concentrations of calcium ions on ice, and then shocked at a higher temperature for a few minutes. Most of the cells die during the treatment, but some survive and take up the DNA. Another method to make E. coli cells competent is to expose them to a high-voltage shock. Electroporation opens the cell wall, allowing the DNA to enter. This method is much faster and more versatile. Electroporation is used for other types of bacteria as well as yeast.

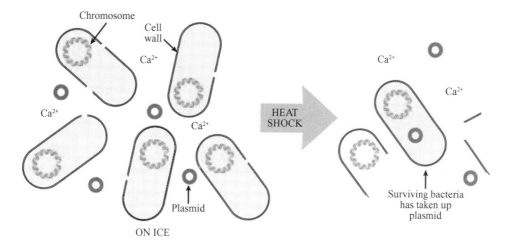

Fig. 3.12 Transformation

Bacterial cells are able to take up recombinant plasmids by incubation in calcium on ice.
After a brief heat shock, some of the bacteria take up the plasmid.

When a mixture of different clones is transformed into bacteria as in a gene library, cells that take up more than one construct usually lose one of them. For example, if genes A and B are both cloned into the same kind of vector and both cloned genes get into the same bacterial cell, the bacteria will lose one plasmid and keep the other. This is due to plasmid incompatibility, which prevents one bacterial cell from harboring two of the same type of plasmid. Incompatibility stems from conflicts between two plasmids with identical or related origins of replication. Only one is allowed to replicate in any given cell. If a researcher wants a cell to have two cloned genes, then two different types of plasmids could be used, or both genes could be put onto the same plasmid.

7 Synthetic Ribozymes Used in Medicine

Ribozymes are beginning to be used in medical applications. Researchers studying AIDS have derived a hammerhead ribozyme that can inhibit HIV replication. This engineered ribozyme was in clinical trials as of 2006. It is administered by expressing the ribozyme gene in a viral vector. The vector is transfected into peripheral blood T lymphocytes from HIV-infected patients. It is hoped that the expressed ribozyme will cleave the RNA version of the HIV genome, thus preventing replication of the HIV virus.

Another ribozyme has been developed to cleave an RNA virus, hepatitis C virus (HCV). HCV is the leading cause of chronic hepatitis, and no vaccine is available. Various engineered ribozymes have been identified that can efficiently cleave HCVRNA, but these studies are still in vitro. The engineered ribozymes have worked efficiently in cell culture where liver cells from infected individuals have been harvested and grown in dishes, but they have not yet been tested directly in patients.

The clinical use of ribozymes has many of the same obstacles as for any new drug. Each new ribozyme must be delivered to the correct location and expressed in cells that are diseased. Each ribozyme must be stable and resistant to degradation. In this regard, many engineered ribozymes contain modified bases, which prevent degradation by cellular endonucleases. Finally, the ribozyme must not have any deleterious side effects. High specificity to their target provides ribozymes with more potential than many preexisting therapies. For example, chemotherapy of cancer patients kills any rapidly dividing cells, not just the cancerous cells. This is why chemotherapy patients lose their hair. Ribozymes recognize one specific target mRNA; therefore, ribozyme treatments may avoid side effects seen in chemotherapy treatments.

Unit 4

··· Part A ···

Food Biotechnology

1 Sweeteners

Sweeteners, sweet substances other than sugar and related carbohydrates, are **polyols** or intense sweeteners. Most of these substances are produced by chemical synthesis. Among the group of polyols, **erythritol** and part of **mannitol** are produced by fermentation. Immobilized cells or enzymes are used in the production of **isomalt** and **maltose**, an intermediate for maltitol. Many papers on the production of sorbitol and **xylitol** by fermentation are available. Among the intense sweeteners, the building blocks of aspartame, aspartic acid and phenylalanine, are produced by fermentation, and enzymatic coupling was used in practice by one producer. **Stevioside** and **glycyrrhizin** can be modified enzymatically, and possibilities to express the genes for **thaumatin**. **Tagatose**, a reduced-calorie carbohydrate, can be produced by enzymatic conversion of **galactose**.

Sweet-tasting substances other than sugar have become increasingly important in food production in the course of the last decades. In certain areas such as soft drinks, the quantity of products sweetened with these substances has almost equalled the conventional, sugar-sweetened products in some countries including the United States. In others, such as in some European countries, the percentage of these beverages has increased steadily after a **harmonized** approval for all Member States of the European Community in 1995. In other fields of application such as **sugar-free** sweets and **confections**, polyols have been established as a **noncariogenic** alternative to sucrose.

The general field of sweet-tasting substances can be divided in two main sectors. One comprises sugar (sucrose) and other nutritive carbohydrates including glucose, fructose, and products obtained from hydrolyzed starch such as high-fructose corn syrup. The other sector covers products generally called sweeteners. They are non-carbohydrate alternatives such as polyols and intense sweeteners. A third group of still rather limited commercial importance comprises sweet carbohydrates of physiological characteristics different from the standard carbohydrates normally used in food production.

New Words

sweetener	(调味用)甜料,甜味剂
polyols	多元醇
erythritol	赤藓糖醇
mannitol	甘露醇,甘露糖醇
isomalt	异麦芽酮糖醇,异麦芽,益寿糖
maltose	麦芽糖
sorbitol	山梨醇,山梨糖醇
xylitol	木糖醇
stevioside	蛇菊苷,甜菊苷(一种非营养性的天然甜味剂)
glycyrrhizin	甘草甜素,甘草酸
thaumatin	奇异果甜蛋白(在热带灌木奇异果的果实中发现的一种甜蛋白)
tagatose	塔格糖(一种已酮糖)
galactose	半乳糖
harmonize	使和谐,使一致
sugar-free	无糖的,未包糖衣的
confection	糖果,零食,小吃,糖果剂
non-cariogenic	不生龋齿的

2 Biopreservatives

The exploitation of **biopreservation** is by no means a new concept. Biotechnological processes for preserving food have already been used for thousands of years, even though the

Unit 4

underlying mechanisms were not understood. Today, biopreservation of foods is as relevant as ever before because it is one of the few possible answers to what at first glance appears to be totally **contradictory** trends and demands:

● Health trends: The levels of salt, sugar and fat in foods are under pressure to be reduced. These changes are beneficial for human health, but they also all confer an increase in water activity, which provides a friendlier environment for microorganisms.

● Taste preferences: In many products, trends are towards a milder (i.e. less acidic) taste, which results in a higher pH that again is less adverse for microorganisms.

● Perception of "natural": This results in milder or minimal processing, which results in a fresher appearance of the food but also less inactivation of unwanted microorganisms. Furthermore, it increases the demand for "preservative-free" products.

● Convenience trends ("practically homemade"): There are two main risks associated with this trend-namely, more extensive processing, which results in more steps in which contamination with detrimental microorganisms can occur, and the need for proper handling by the consumer (e.g. sufficient heating), which may be neglected.

● Durability and open shelf-life: Market access and economically viable logistics require a long shelf-life. Furthermore, a sufficient open shelf-life is required to ensure customer loyalty.

● Ethical issues: Concerns such as corporate social responsibility, carbon dioxide (CO_2) footprint, and fair-trade and organic products put restrictions on which solutions a food producer can employ.

All in all, these trends lead tofood formulations that provide better growth conditions for microorganisms, milder processing that results in less initial reduction, more processing steps that increase the risk of contamination, a need for longer shelf-life, and pressure to reduce food waste. In addition, many of the conventional preservatives are deemed to be unacceptable by trendsetters and consumers. Everyone wants preservative-free food, but most will agree that we cannot maintain our present society and standard of living-and certainly cannot reduce the global food waste problems-with food that is not preserved.

New Words

biopreservation	生物保鲜
contradictory	矛盾的,反对的,反驳的,抗辩的
inactivation	[生物]失活,[化学]钝化(作用)

preservative-free	无防腐剂的
homemade	自制的,家里做的
shelf-life	货架期,保质期
logistics	物流,后勤
food formulations	食品配方
trendsetter	潮人,领导流行的人

3 Nisin

Nisin is a **cationic**, **amphiphilic** peptide produced by various strains of *Lactococcus lactis*, which has a relatively broad target spectrum that inhibits a wide range of **gram-positive bacteria**. The **antimicrobial** property of nisin was first observed in 1928, when it was reported that inhibition of a dairy **starter culture** was caused not by phages but by a strain of *L. lactis* (formerly called lactic streptococci and group N streptococci). The inhibitory compound was further studied the following years and given the name nisin, alluding to its origin as a "group N streptococci Inhibitory Substance". The application of nisin for preservation of dairy products was suggested already in 1951 for inhibiting blowing of **Swiss-type cheese**. Soon after, the first commercial preparation was made by Aplin and Barret in 1953. The use of nisin as a food preservative was approved by the Food and Agriculture Organization of the United Nations and World Health Organization (WHO) in 1969, by the European Union (EU) in 1983 (E 234), and granted Generally Recognized As Safe (GRAS) status by the U.S. Food and Drug Administration (FDA) in 1988.

Thus, nisin has a long history of safe use in food. It is the only purified **bacteriocin** that is widely approved as food additive-a fact which presumably also reflects its early discovery. Through the years, a substantial number of scientific papers have described the biosynthesis, chemical and physical properties, mode of action, and practical applications of nisin. A short summary is given below, with focus on aspects that affect industrial applications.

Nisin belongs to the **lanthionine**-containing bacteriocins. Production of bacteriocins containing the unusual lanthionine residues, which are formed by **posttranslational** modifications, is not uncommon amongst **lactic acid** bacteria; linear, globular, and two-peptide variants have been characterized. Many of these peptides are effective at low concentrations against a wide range of gram-positive bacteria, which has been attributed to a common mode of action: nisin and other lantibiotics bind with high affinity to a docking molecule in the cell envelope of target bacteria,

Unit 4

lipid II, an intermediary molecule for building bacterial cell walls. Nisin is a linear lantibiotic that exerts its antibacterial action through inhibiting cell wall formation as well as forming **membrane pores**; it is furthermore active against **spores**. Several variants of nisin occur naturally; the two that are currently available as commercial products, nisin A and nisin Z, differ in one amino acid, which confers a difference in charge and solubility.

Nisin was first approved for food applications in 1969. The initial approvals were based on **toxicity** testing results from 1962. Recently, two independent studies have shown that even a very high daily intake was not toxic. In the EU, nisin is currently approved as additive in **ripened** and processed cheese, **clotted** cream, puddings such as **semolina** or **tapioca**, **mascarpone**, and pasteurized liquid egg. In the United States and Australia/New Zealand, further approvals have been granted, such as for use in sauces, soups, salads, **dressings**, and ready-to-eat and processed meat products.

New Words

nisin	乳酸链球菌肽；尼生素
cationic	阳离子的
amphiphilic	两亲的，两性分子的
gram-positive bacteria	革兰氏阳性菌
antimicrobial	抗菌的
starter culture	发酵剂，起子培养
streptococci	链球菌(streptococcus 的复数)
Swiss-type cheese	瑞士干酪
bacteriocin	细菌素，球菌素，乳酸菌肽
lanthionine	羊毛硫氨酸
posttranslational	转译后的
lactic acid	乳酸
membrane pores	膜孔
spores	孢子(spore 的复数)
toxicity	[毒物] 毒性
ripened	成熟的
clotted	凝结的
semolina	粗粒小麦粉，粗面粉

tapioca 木薯淀粉
mascarpone 马斯卡泊尼乳酪
dressings 敷料剂；调味品

4 Colorants

The color of food and drinks is important, as it is associated with freshness and taste. Despite that natural **colorants** are more expensive to produce, less stable to heat and light, and less consistent in color range, natural colorants have been gaining market share in recent years. The background is that artificial colorants are often associated with negative health aspects. Considerable progress has been made towards the fermentative production of some colorants. Because colorant biosynthesis is under close metabolic control, extensive strain and process development are needed in order to establish an economical production process. Another approach is the synthesis of colors by means of biotransformation of adequate **precursors**. **Algae** represent a promising group of microorganisms that have shown a high potential for the production of different colorants, and dedicated fermentation and downstream technologies have been developed.

Colorants have been added to food and drinks for a long time. Important functions of food and beverage colors are the identification and recognition of **spoilage** that are associated with the food or beverage as such, but also with its taste and smell. Colorants are added in order to emphasize and to keep the original color after manufacturing for prolonged periods and for **perception** of freshness and shelf life; color makes the food more attractive for the consumer. For example, many people associate **curry** with its specific taste as the typical color and taste of curry are closely related. This is also the case for **saffron**. Furthermore, it is hardly possible to obtain the bright colors of soft drinks without the addition of colorants. Most of the natural dyes are obtained from plants, and the most commonly applied are **β-carotene**, **lycopene**, and **chlorophyll**. Some of them may have favorable health properties, for example, **antioxidant** or even **anticarcinogenic** effects.

A very old application of natural colors has been reported in Japanese **Shosoin** texts, the use of natural colorants for the **colorization** of soy beans and **adzuki** was described (eighth century). In the nineteenth century, many natural colorants were replaced by synthetic dyes because of lower production costs, higher stability, and a more consistent color range. The new colorants have passed an extensive **trajectory** of tests, and were registered in Europe before they were allowed for use in food and drinks.

Despite the fact that natural colorants are typically more expensive to produce, less stable to heat and light, and less consistent in color range, natural **pigments** have been gaining market share as food and feed colorants and **nutraceuticals** in recent years. Consumers increasingly prefer natural colorants as synthetic colorants are associated with allergenic reactions, hyperactivity, and even bad taste. Furthermore, more and more evidence appears in the scientific literature about undesired and potential toxicological effects of synthetic colorants. This prompts the regulatory authorities to shorten the list of permitted synthetic food colorants drastically. Furthermore, the FAO/WHO Expert Committee on Food Additives advised the further removal of small chemical impurities from food colors.

New Words

colorant	色素,着色剂,染色剂
precursor	前体（化学）,先驱,前导
algae	藻类,海藻
spoilage	腐败,损坏,糟蹋,掠夺,损坏物
perception	知觉,感觉,看法,洞察力
curry	咖喱粉,咖喱,咖喱饭菜
saffron	藏红花,番红花,橙黄色
β-carotene	β-胡萝卜素
lycopene	番茄红素
chlorophyll	叶绿素
antioxidant	抗氧化剂,硬化防止剂,防老化剂
anticarcinogenic	抗癌的;抑制癌发生的
Shosoin	正仓院（日本奈良时代的仓库,位于日本奈良市东大寺内）
colorization	着色,颜色迁移
adzuki	红豆,小豆
trajectory	轨道,轨迹
pigment	颜料,染料,色素
nutraceutical	保健品,营养药品,营养物质,营养素

5 Acidic Organic

Some of the most important and frequently used additives in beverage, food, and feed production are organic acids and their derivatives. Organic acids are acidic and contain carbon atoms. Often, they are products of metabolism. Therefore, many of these acids are advantageously produced via biotechnology.

Organic acids could be divided into several groups. The most common group comprises the **carboxylic acids**, which contain one or more **carboxyl groups** (—COOH). Important examples, especially for beverage, food, and feed applications, are acetic acid (one carboxyl group), **malic acid** (two carboxyl groups), and citric acid (three carboxyl groups). Organic acids are often weak acids that act as buffers in aqueous solutions. Buffering capacity is particularly interesting for beverage, food, and feed production. Organic compounds containing the functional group—SO_2OH (**sulfonic acids**) are somewhat stronger acids than the carboxylic acids. One natural occurring example is **taurine**, which is added to dry food for cats. Moreover, other functional groups such as **alcohol-**, **thiol-**, **enol-** and **phenol** groups can be responsible for a certain acidity of an organic compound. These substances are in general weak acids.

Organic acids can also be classified according to their occurrence in the **metabolism** of organisms. Some organic acids are part of the central metabolism that is essential for the energy supply of cells. These compounds have a low molecular weight and can often be produced at high titer by fermentation of microorganisms. Acidic additives produced in the primary metabolism are therefore frequently used in food, feed, and beverage manufacture.

Only a few organic acids belonging to secondary metabolic processes are used in beverage, food, and feed production. In general, these have special properties (e.g. **ferulic acid and lactobionic acid**) or could be easily produced by chemical synthesis (e.g. **benzoic acid**).

Most organic acids and their **derivatives** could in principle be produced with biotechnological methods. Unfortunately, not all of them are produced in an economically sustainable manner due to higher costs of such bioprocesses compared to synthetic production or lack of knowledge regarding biotechnological routes and downstream processing.

Unit 4

New Words

carboxylic acid	羧酸
carboxyl group	羧基
malic acid	苹果酸;羟基丁二酸
sulfonic acid	硫酸
taurine	牛磺酸;氨基乙磺酸
alcohol-	醇
thiol-	硫醇
enol-	烯醇
phenol	苯酚
metabolism	代谢,新陈代谢
ferulic acid	阿魏酸
lactobionic acid	乳糖醛酸,乳糖酸
benzoic acid	苯甲酸
derivative	衍生物,派生物

Part B

Production of Food and Feed Additives

1 β-Carotene

β-Carotene is applied as a food coloring additive, as provitamin A in food and feed, in multivitamin products, as an antioxidant, and as a colorant for cosmetics. In the literature, a number of microorganisms are described that are capable of accumulating β-carotene up to relatively high levels, for example, the fungi *B. trispora*, *Phycomyces blakesleanus*, the yeast *Rhodotorula glutinis*, and the above-mentioned alga *D. salina*. Production processes were developed with mutants of *B. trispora* in the former Soviet Union and for *D. salina*.

Among the β-carotene-producing algae, *D. salina* is currently the best-known and most efficient producer: *D. salina* cells are able to synthesize β-carotene in levels a thousand times higher than those of carrots. *Dunaliella* is also able to synthesize α-carotene, violaxanthin, neoxanthin, zeaxanthin, and lutein. The unicellular green alga *D. salina* is grown in open ponds without (hardly any) process control. The production facilities are located in places where optimum conditions are found, namely in Australia, Israel, and in the United States; these places benefit from a lot of sunshine, little cloudiness, the availability of saline water, and high average temperatures.

In Australia, β-carotene production is performed in an extensive manner. The ponds have a large surface, and there are no measures for active mixing or other ways of control. Therefore, mixing only occurs by wind, convection, and diffusion, and the biomass and β-carotene yields are low (0.1g β-carotene/m^3). The effect of the seasons on the production conditions is low, and this enables year-round production. The light intensities are optimal for maximum carotenoid synthesis. In order to maintain a stable production population without too much pressure from competitor and predator organisms, the salt concentration is kept under close control. At a salinity of 12% salt

Unit 4

(w/v), the formation of biomass is optimal, but the β-carotene synthesis is low. In addition, these conditions are favorable for the development of undesired organisms, whereas at salinities of 24% ~ 27% most favorable production conditions are realized in terms of β-carotene production, biomass formation, and culture stability.

❷ Natamycin

Natamycin was discovered in the 1950s. As described by Struyk et al., "A new crystalline antibiotic, pimaricin, has been isolated from fermentation broth of a culture of a *Streptomyces* species, isolated from a soil sample obtained near Pietermaritzburg, State of Natal, Union of South Africa. This organism has been named *Strepyomyces natalensis*". The original name "pimaracin" can be found in earlier publications but it is no longer accepted by the WHO. Natamycin is classified as a macrolide polyene antifungal and is characterized by a macrocyclic lactone-ring with a number of conjugated carbon-carbon double bonds (Fig. 4.1).

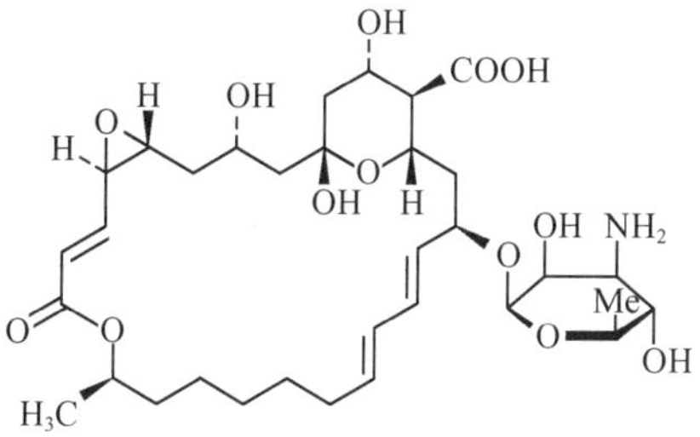

Fig. 4.1 The chemical structure of natamycin

Natamycin has a low solubility in water (approximately 40 ppm (ppm = 10^{-6})), but the activity of neutral aqueous suspensions is very stable. Natamycin is stable to heat and it is reported that heating processes for several hours at 100 ℃ lead to only slight activity losses. Natamycin is active against almost all foodborne yeasts and molds but has no effect on bacteria or viruses. The sensitivity to natamycin in vitro (minimal inhibitory concentration) is in most cases below 20 ppm.

Natamycin acts by binding irreversibly with ergosterol and other sterols, which are present in the cell membranes of yeasts and vegetative mycelium of molds. It disrupts the cell membrane and increases the cell permeability, which finally leads to cell death. The fungicidal of natamycin is an "all-or-none" effect, which destroys the cell membrane of the target cells.

Due its interaction with ergosterol, which is a major constituent of fungal cells, it is unlikely that fungi will develop resistance. So far, after many decades of use, no development of resistance has been reported. Natamycin is mostly used for surface applications, particularly for treating surfaces of hard cheese and salami-type sausages. One of the advantages over sorbate is that even the dissolved. Natamycin can be applied by spraying the surface (e. g. of cheese), by dipping, by applying natamycin via coating emulsions or by direct addition.

Commercial preparations are produced by fermentation of sugar-based substrates by *Streptomyces natalensis*. Natamycin is then recovered by extraction, filtration, and spray drying. The dried powder can be stored for years without any activity loss.

3 Fermentative Production of Food Grade Colorants

In nature, many microorganisms are found that are able to synthesize colored metabolites, for example, chlorophylls, carotenoids, melanins, flavins, violacein, and indigotin. This opens up opportunities for their production by means of biotechnology.

There are a number of examples of "biocolorants" that areeconomically interesting to produce by fermentation, for example, saffron. Saffron is present in the stigma of Crocus sativa, albeit in high levels, but only comprises a very small part of the total plant. In order to meet the demand, a significant part of the agricultural area is necessary (Iran: 80,000 hectares), and harvesting and processing of the stigmas are very laborious. Therefore, studies were performed to grow saffron-producing tissues in vitro. It was technically possible, but the yields were far away from economic requirements.

Therefore, it is recommended to investigate microorganisms that are easy to cultivate for their ability to produce colorants. Until now, only some of the colorants produced by plants have been known to be produced in microorganisms. If a suitable microorganism is not available yet, a screening program is an alternative. With regard to this, algae represent an interesting group: the interest in these autotrophic or facultative heterotrophic microorganisms is strongly increasing as some species produce lipids up to very high levels (30%), which may contribute to a more sustainable fuel supply. There is currently a strong focus on the development of fermentation technologies that allow for economical production. Autotrophic algae need light for their energy supply, and therefore specific fermentation equipment. Furthermore, harvest and downstream processing steps have to be optimized. These cost-reduction programs for biofuel production may also be helpful for the production of colors by algae, for example, of astaxanthin, β-carotene, and lutein.

Unit 4

4 Dough Conditioner

Dough conditioners, also known as flour treatment agents or improvement agents, are frequently used in bakeries. Organic acids have two functions in flour improvement. They are added for oxidation, leading to maturing of the flour and thus improving the baking quality. Secondly, organic acids are added for their reducing character in continuous dough mixing. Reduction reactions enhance the effectiveness of mixing. The mixing time to achieve proper dough development is decreased at a given mixing speed. The needed energy input for dough mixing could be reduced.

5 Leavening Agent

Chemical leavening is very common in the production of bakery products. By producing gas through a chemical reaction, the dough rises and becomes fluffy, which is desired for most bakery products. Usually, sodium bicarbonate (bakery soda) is used for this reaction. If bakery soda is heated, sodium carbonate is formed by releasing carbon dioxide and water. However, the sodium carbonate is unwanted in many applications because of its bitter taste and a tendency to produce yellowish color. A possible solution is the addition of a weak acid. Then, in a first step, the sodium bicarbonate reacts with the acid. Carbonic acid and a sodium salt of the acid are produced. Afterwards, carbonic acid decomposes to water and carbon dioxide. Fig. 4.2 shows the chemical leavening reaction using the example of acetic acid as an additive. Important organic acidic sources for chemical leavening with sodium carbonate are vinegar (acetic acid), lemon juice (citric acid), molasses or buttermilk (lactic acid) and cream of tartar (potassium bitartrate). The correct mixture of soda and acid has to be used for a good product. If there is not enough acid, the product turns yellowish and bitter. Excess acid produces a sour taste. Commercial baking powder is an optimized mixture of sodium bicarbonate, an organic acid and a dry diluent (e.g. corn starch).

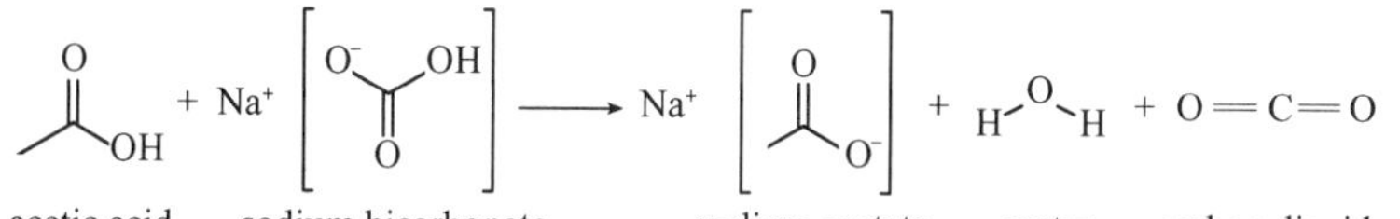

Fig. 4.2 Example for a chemical leavening reaction

6 Function of Acidifiers in Animal Feed

Animal feed is produced, transported, and stored in large quantities. Therefore, a certain level of contamination with unwanted microorganisms is essentially unavoidable. The level of microbes could rise rapidly under favorable temperatures and moisture conditions. Contamination reduces the nutritional value of the feed. Additionally, certain pathogens (e.g. Salmonella) could be dangerous for the animals, even at low titers. By carryover, pathogens or their produced toxins may also enter the human food chain. Hence, feed is often treated with heat and acidic substances are added. The preservation mechanism is the same as in food. The heat reduces the potential of initial contamination and microbial growth is inhibited by reducing the pH of the feed. Moreover, different acidic compounds act specifically against certain microorganism groups. Thus, the uptake of pathogens and toxins by farm animals with feed is minimized.

The addition of acidifiers into diets has positive effects for the health of the animals and the growth performance of farm animals can be enhanced. Several mechanisms describing the reasons of these beneficial effects have been proposed. First, certain organic acids and their derivatives have antimicrobial activities as described before. Pathogens may be reduced in animals fed using feed with acidifiers. The gastric pH could be reduced by acidifiers. A low pH in the stomach promotes the activity of pepsin, which improves the protein digestion in swine, for example. The reduction of the gastric pH is somewhat controversial. Further studies have shown that the influence of organic acids in this case depends on the chosen acidifier. The chelating function of some organic acids and their derivatives has a positive effect on animal growth.

Often mixtures of acidifiers are used to maximize the beneficial effects. Due to the positive effect on health of the animals, acidifiers are increasingly accepted as an alternative to antibiotics.

Additionally, acidifiers enhance the nutritional value of the feed, which leads to a better growth of the animals. Most organic acids and their derivatives have high energy contents (e.g. heat of consumption for propionic acid: 4,968 kcal/kg). This has to be taken into account when planning the feed rations to avoid overfeeding, if organic acids are added.

Unit 5

Part A

Solid-state Fermentation

1 Introduction

After the 1940s, the huge demand for acetone, **butanol**, and **penicillin** caused the liquid fermentation industry to increase rapidly. The **solid-state fermentation** industry began to decline, and the advantages of solid-state fermentation were concealed by the rapid development of **liquid fermentation**. Consequently, solid-state fermentation only constitutes a small part of the fermentation industry as a whole. Currently, the problems of liquid fermentation, such as high energy consumption and serious pollution, are becoming increasingly prominent, significantly limiting the sustainable development of fermentation. People are again taking note of the advantages of solid-state fermentation, such as its water-saving, energy-saving, and low-cost properties. Solid-state fermentation has begun to play an important role in the chemical, **pharmaceutical**, and environmental fields, which points out a clear direction for the sustainable development of the entire biological and chemical industry. Thus, the principles and applications of solid-state fermentation have become a new research hot spot in recent years.

Fermentation is the process by which microorganisms catalyze nutrients, synthesize **secondary metabolites**, and complete other physiological activities under **anaerobic** or **aerobic** conditions. During the process, the desired microorganisms or microbial metabolites are accumulated. Therefore, there are three elements of fermentation research: the clear target product, the producing strain, and the desired training environment (nutrients, temperature, humidity, oxygen, etc.).

The unique feature of solid-state fermentation is that there is nearly no free water in the solid substrate. Solid-state fermentation is a three-phase system consisting of the continuous gas phase, the liquid film, and the solid phase. It should be noted that there is no obvious relationship between the water content of the substrate and the content of free water. Because of the strong **hydraulic** holding ability of the solid substrate, such as that of the sugar beet plant material, even if the water content of the substrate is more than 80%, there is seldom free water among the solid substrate. Consequently, the content of water cannot be defined as the only standard of solid fermentation.

The substrate can be divided into two categories based on its digestibility: nutritional carrier substrate or inert carrier substrate. The nutritional carrier substrate is crops (wheat bran, soybean meal, etc.) or agricultural and forestry wastes (straw, bagasse, sawdust, etc.). This substrate not only performs as a physical structure for the growth of microbes but also provides a carbon source, a nitrogen source, and growth factors for the microorganisms. Nutritional food crops and agricultural and forestry wastes are the most commonly used substrates in actual production applications. The inert carrier substrate is a porous substrate that is chemically inert and difficult to be decomposed by microorganisms, such **aspolyurethane foam**, **macroporous resin**, **perlite**, and **vermiculite**. These substrates only play a supporting role in the fermentation process; microorganisms can obtain nutrition from the fluid culture that is distributed in the porous media gap.

Tab. 5.1 presents the development process of solid-state fermentation. From this, we can conclude that solid-state fermentation technology has provided many products for humans since the beginning of human civilization. In recent years, the application of solid-state fermentation technology has greatly expanded.

Tab.5.1 Development of solid-state fermentation products

Time	Products
2000 B.C.	Bread, vinegar
1000 B.C.	Sauce, koji
550 B.C.	Kojic acid
7th century	Kojic acid was introduced to japan
Sixteenth century	Tea

Unit 5

Tab. 5.1 (Continued)

Time	Products
Eighteenth century	Vinegar
1860—1900	Sewage treatment
1900—1920	Enzyme
1920—1940	Gluconic acid, citric acid
1940—1950	Penicillin
1950—1960	Steroid
1960—1980	Protein feed
1990	Bioremediation, biological detoxification, biotransformation, biopulping, aflatoxin, ochratoxin, endotoxin, gibberellic acid, zearalenone, cephamycin

New Words

butanol	丁醇,正丁醇
penicillin	盘尼西林(青霉素)
solid-state fermentation	固态发酵,固体发酵
liquid fermentation	液体发酵,液态发酵,液体深层发酵
pharmaceutical	药物,制药学
secondary metabolite	次级代谢,次级代谢物
anaerobic	厌氧的
aerobic	需氧的,好氧的
hydraulic	液压的,水力的,水力学的
substrate	培养基,酶作用物,底物
polyurethane foam	聚氨酯泡沫塑料,泡沫聚氨酯
macroporous resin	大孔树脂,大孔吸附树脂
perlite	珍珠岩
vermiculite	蛭石(一种隔热材料)

2 Advantages and Applications of Solid-State Fermentation

The water content of the solid substrate can be effectively maintained in the range of 12% - 80%, mostly around 60%. In contrast to solid-state fermentation, the typical water content of liquid fermentation is above 95%. The current fermentation technology is liquid fermentation. Although application of and research for this technology have been long term, it still has many problems that need to be overcome.

Compared to liquid fermentation, the main advantage of solid-state fermentation is a sufficient supply of **oxygen**. There is less **organic** wastewater and higher product yield in solid-state fermentation. The solid environment is more similar to the natural habitat of **filamentous fungi**. High value-added products could be produced by solid-state fermentation using low-cost industrial and **agricultural residues** as substrate. Consequently, solid-state fermentation is the most **promising** technology that can **comprehensively** utilize **renewable resources**.

At present, solid-state fermentation products mainly include traditional foods (vinegar, soy sauce, flavor spices); microbial cells (single-cell protein, **spirulina**, **edible fungus**, etc.); microbial enzymes (**amylase**, **glucosidase**, **cellulase**); and other microbial metabolites (**nucleotides**, **lipids**, **vitamins**, amino acids, etc.).

New Words

maintain	维持,保持,继续
oxygen	氧,氧气
organic	有机的,组织的,器官的
filamentous fungi	丝状真菌
agricultural residue	农业废弃物,农作物废弃物
promising	有希望的,有前途的
comprehensively	综合的,全面的
renewable resource	可再生资源,可再生能源
spirulina	螺旋藻,螺旋藻属
edible fungus	食用菌,木耳
amylase	淀粉酶

glucosidase	葡萄糖苷酶
cellulose	纤维素
nucleotide	核苷,核苷酸
lipids	脂肪,脂类
vitamin	维生素

 Aerobic Solid-State Fermentation

Oxygen is one important factor that affects the process of aerobic solid-state fermentation. Based on the nature of biological processes, aerobic solid fermentation can be defined as a biological metabolic process that uses air containing oxygen as the continuous phase. Solid-state fermentation involves the growth of microorganisms on moist solid particles. There is a continuous gas phase in the space between the particles. The majority of water of the system is absorbed within the moist solid particles, and there are thin water films on the particle surfaces. The **interparticle water phase** is discontinuous, and most of the interparticle space is filled by the gas phase. In the natural environment, the majority of microorganisms live under aerobic conditions, so the aerobic solid fermentation processes simulate the natural environment, and they may be more suitable for the growth of microorganisms.

With regard to solid-state fermentation equipment, researchers have developed **tray-type bioreactors**, **packed bed bioreactors**, **rotating drum bioreactors**, gas-solid fluidized bed bioreactors, and gas double **dynamic** bioreactors. In 1929, the British **scholar** Fleming first discovered that bacteria could not grow in the plate where Penicillium had grown and named this antibacterial substance penicillin. This began the **era** of large-scale study and use of **antibiotics**. In the initial stage, penicillin was produced using aerobic tray fermentation. Because of the limitations of the production process, the levels of production, extraction, and purification were low. In the 1940s, with the development of submerged liquid fermentation technology, the production of penicillin was scaled up to the industrial level, which opened a new chapter of modern aerobic fermentation.

For different products or fermentation technologies, the processes of an aerobic solid-state fermentation procedure may be different, but the basic properties can be **summarized** in the following aspects:(1) **Pretreatment** of raw materials, such as crushing, cooking, molding, **starter propagation**, cooling, and so on. (2) Compared to the liquid fermentation process, the flow properties of the solid substrate are poor. Consequently, material handling is an important factor that influences the efficiency of the solid-state fermentation process and should be paid more attention. (3) Microorganisms in aerobic solid-state fermentation include some natural microorganisms and some artificial **screening** strains. (4) For the process and control of solid-state fermentation with respect to liquid fermentation, the solid substrate environmental conditions are more complex, and the fermentation process is more difficult to control. (5) Compared to anaerobic solid fermentation, besides the transfer of mass and heat, the distribution and transfer of oxygen in a fermentor are other important factors that influence the fermentation process. (6) Solid-state fermentation postprocessing consists of product purification, product drying, sterilization, repackaging, and so on.

New Words

interparticle	粒子间的,颗粒间的
waterphase	水相
tray-type bioreactor	盘式生物反应器
packed bed bioreactor	填充床生物反应器
rotating drum bioreactor	转鼓式生物反应器
dynamic	动态的,动力的,动力学的,有活力的
scholar	学者
era	时代,年代,纪元
antibiotics	抗生素,抗菌药物,抗菌物质
summarize	总结,概述
pretreatment	预处理,前处理
starter propagation	制曲;发酵剂扩大培养
screening	筛选的
fermentor	发酵罐,发酵器
postprocessing	后加工,后部工艺
sterilization	消毒,杀菌,灭菌

❹ Anaerobic Solid-state Fermentation

The fermentation industry originated in China. A large-scale fermentation industrial system has been established in recent decades. The essence of the fermentation industry is the **deep processing** industry for agricultural products, which is an extension of the industrialization of agriculture. But, it causes problems, such as serious pollution and high energy consumption in modern industry. The main environmental pollution caused by the fermentation industry is water pollution, mostly from the residue after processing of the raw materials, such as **bagasse** or **beet pulp**. The waste liquor is also from separation and extraction of products, such as waste liquor in **monosodium glutamate** fermentation and the washing and cooling water from the production process. Although the fermentation industry has made great progress in cleaner production and pollution prevention, because of rapid growth in the yield of fermentation products in China, wastewater and the total emission of pollutants still evidence a growth trend. The outstanding problems of the fermentation industry are still how to comprehensively utilize resources; solve the problems of grain saving, energy saving, water saving, and environmental pollution; and realize clean production.

Solid-state fermentation is the fermentation process completed by one or more microbes on a wet solid-state substrate with little or no free flow of water. From the nature of the biological reactions, solid-state fermentation is the **continuous phase bioreactor** process. The water content of the solid substrate can be effectively controlled at between 12% and 80%.

Anaerobic fermentation is carried out in **sealed** conditions and does not need to **aerate** in confined conditions. The fermentation equipment is relatively simple and has low energy consumption.

In short, anaerobic solid-state fermentation has the unique advantages in that it is water saving and energy saving and protects the environment; it will be the future direction of the fermentation industry. People need to recognize anaerobic solid-state fermentation to guide cleaner production by the fermentation industry.

Solid-state fermentation can be divided into anaerobic and aerobic solid-state fermentation, depending on oxygen utilization. The pathway of microbial **carbohydrate** metabolism in anaerobic conditions is different from that in aerobic conditions, as is the product obtained. In addition, there are still many differences between anaerobic and aerobic solid-state fermentation (Tab. 5.2).

Tab. 5.2 Comparison of anaerobic solid-state fermentation and aerobic solid-state fermentation

Species	Aerobic solid-state fermentation	Anaerobic solid-state fermentation
Fermentation conditions	Maintain ventilation oxygen; strict control of temperature and humidity in the gas supply	Without ventilation (the strict anaerobic fermentation required to drive oxygen), but requires large doses of inoculation
Fermentation microorganisms	Most aerobic bacteria; a wider range of bacteria sources	Usually anaerobes or facultative Anaerobes
Fermentation characteristics	Microbial growth fast; short fermentation period	Poor growth microorganisms; long fermentation period; can form the unique flavor of the product
Application	Enzymes, antibiotics, etc	Liquor, biogas, and fuel ethanol

Anaerobic solid-state fermentation can facilitate the microbes to produce rich flavor substances. Many of the fermentation processes are a combination of aerobic and anaerobic fermentation, such as traditional soy sauce brewing. It is **noteworthy** that anaerobic solid-state fermentation provides a unique flavor that is **irreplaceable** in food fermentation and liquor production. Anaerobic solid-state fermentation has made full use of this advantage for developing fermented products with more flavor variety that is rich in nutrients.

New Words

deep processing	深加工, 深度处理
bagasse	甘蔗渣
beet pulp	甜菜浆, 甜菜粕
monosodium glutamate	味精(主要成分是谷氨酸一钠)
continuous phase bioreactor	连续相生物反应器
sealed	密封的, 密闭的
aerate	通气, 充气, 让空气进入, 使曝露于空气中
carbohydrate	碳水化合物, 糖类
noteworthy	值得注意的, 显著的
irreplaceable	不能替代的, 无可取代的, 不能调换的

Unit 5

··· Part B ···

Applications of Solid-state Fermentation

 **1** Application in Silage

Straw resources as a feed source are extremely abundant, which can save much food and indirectly provide animal protein products for human beings. Therefore, it is the key point of the scale development and utilization of straw as feed resources by scientific pretreatment technology, such as anaerobic solid-state fermentation, to improve the nutritional value and palatability of straw feed as raw material. Solid state fermentation using straw to produce feed is a process that uses the acid produced by lactic acid bacteria to reduce the pH value of feed and inhibit the growth and reproduction of harmful microorganisms. At the same time, the straw can be transformed into good quality feed with higher nutritional value and palatability, such as silage and microstorage of feed.

Fresh silage material is put into a sealed silage container; plant cells do not die immediately, but in 1 – 3 days, and the organic matter is decomposed into acid through their respiration. When the oxygen is depleted, the anaerobic solid-state fermentation starts. At the beginning of silage, aerobic microorganisms, such as yeasts, molds, and lactic acid bacteria, can attach to the raw materials and rapidly reproduce using soluble carbohydrate. The oxygen is quickly depleted by the activity of aerobic microbes and respiration of plant cells, which results in the formation of an anaerobic environment. In addition, heat is generated by the respiration of plant cells, enzymatic oxidation, and microbial activities. The anaerobic and warm environment creates suitable conditions for lactic acid fermentation. The stage is a necessary process for the formation of an anaerobic environment from the aerobic environment in silage.

The aerobic respiration stage should be as short as possible. If the oxygen content in the silage is too high, the plant respiration time is too long, and aerobic microbial activity is too vigorous, the temperature of the raw materials will rise. Sometimes, the temperature can reach up to

60 ℃, thus weakening the competitiveness of the lactic acid bacteria and other microorganisms. The silage nutrients are lost, and quality declines excessively.

With the completion of the period of anaerobic fermentation, small amounts of *Lactobacillus* begin to reproduce rapidly in the plant material. The carbohydrates can be converted into lactic acid by *Lactobacillus*, and the pH value of the crop straw can be reduced by the formation of a large amount of lactic acid, which results in the inhibition of bacterial growth and enzyme activity and the promotion of aerobic microorganism activities. When the pH is decreased to below 4.2, a variety of harmful microorganisms cannot survive, and even the *Lactococcus lactis* activities are also suppressed. When the pH is decreased to 3 continuously, the activity of *Lactobacillus* is inhibited, which is basically the end of the lactic acid fermentation process.

2 Alkaline Protease Production by SSF

Proteases are by far the most important group of enzymes produced commercially and are used in the detergent, protein, brewing, meat, photographic, leather, and dairy industries. These enzymes offer advantages over the use of conventional chemical catalysts for numerous reasons; for example, they exhibit high catalytic activity and a high degree of substrate specificity, can be produced in large amounts, and are economically viable. The detergent industry has now emerged as the single major consumer of several hydrolytic enzymes acting at highly alkaline pH. The major use of detergent-compatible proteases is in laundry detergent formulations.

Until now, alkaline protease has been produced mostly by SmF, a process plagued with problems, including serious pollution, low product concentration, and high production cost. Compared to SmF, SSF has received more attention recently as it uses simpler fermentation medium, requires a smaller space, is easier to aerate, and has higher productivity, lower waste water output, lower energy requirement, and less bacterial contamination. Chen and Wang evaluated the potential of Bacillus pumilus AS 1.162 5 for producing alkaline protease by SSF on PUF.

3 Bacterial Cellulose Production by SSF on PUF

Bacterial cellulose is an extracellular product of vinegar bacteria. It is almost pure cellulose and contains no lignin or other foreign substances. It is the same as cell wall cellulose according to

Unit 5

its chemical composition and reactivity.

Bacterial cellulose is less than 130 nm wide; it is composed of a bundle of much finer microfibrils, less than 10 nm diameter. Bacterial cellulose belongs crystallographically to cellulose I, commonly with natural cellulose of vegetable origin, in which two cellobiose units are arranged parallel in a unit cell, and those cellulose molecules tend to have a specific planar orientation in dried film. The long-chain molecules are aligned parallel in the extended form.

Compared with plant cellulose, bacterial cellulose has some advantages, such as high purity, degree of polymerization, crystallinity, hydrophilicity, Young's modulus, strength, and so on. Bacterial cellulose also has high biocompatibility and good biodegradability. Therefore, it has extensive applications. Among various applications studied so far, it has reached the level of practical use for acoustic diaphragms as bacterial cellulose has been found to bear two essential properties: high sonic velocity and low dynamic loss. The use of films as the raw material for conductive carbon film was investigated and found excellent, although this remains in the laboratory. The use of fragmented bacterial cellulose for papermaking is promising, and test pieces of flexure-durable papers and papers with high filler content, ideal for banknotes and bibles, have been prepared. Other ideas include the use of sheet or film as a temporary skin for medical care and as a separation membrane, the use of fragmented suspension as a viscosity-enhancing agent for various purposes, and others.

The species of bacteria that produces cellulose is generally called *Acetobacter xylinum*. Culture is carried out normally in a static condition at around 28-30℃. The system becomes turbid, and after a while, a white pellicle appears on the surface; its thickness increases steadily with time, reaching over 25 mm in 4 weeks. It is important to note that in the process of gel growth, the aerobic bacteria generate cellulose only in the vicinity of the surface, so that productivity depends primarily on surface area, not on vessel volume. With the aim of enhancing productivity, culture in agitated conditions has been studied, although a flat gel is no longer obtained, and use has to be limited to such applications as papermaking.

4 Biomass Bioconversion Technology Based on Solid-state Fermentation

Based on the analysis of solid material characteristics, the applicability of solid biomass in SSF, and the research study of pretreatment, component fractionation, and fractional conversion, I propose a value-added bioconversion system for biomass focused on SSF in this chapter (Fig. 5.1). This system includes four parts according to the purpose of SSF technology used: biological pretreatment technology, enzyme production for biomass bioconversion, high value-added bioconversion of biomass, and low value-added bioconversion. This section introduces this system and corresponding advances achieved in recent years, especially progress achieved in my work group.

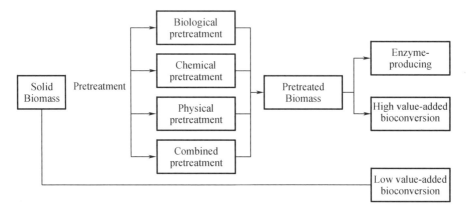

Fig. 5.1　Bioconversion technology for biomass based on solid-state fermentation

5　Solid-state Fermentation Process Control Parameters

The solid-state fermentation control process parameters are closely related to the metabolic regulation of microorganisms. Based on the metabolic needs of the fermentation microorganisms, the control of water activity, oxygen content, temperature, and pH are the main solid-state fermentation parameters. In the solid-state fermentation process, the water, gas, and heat caused by the growth microbes are the dominant factors that determine the environmental changes. The growth and metabolism of the microorganisms are affected by the mass and heat transfer properties of the substrate itself; at the same time, the growth of microorganisms and the utilization of nutritional carrier substrate could change the structure and physical properties of the substrate (Fig. 5.2).

Unit 5

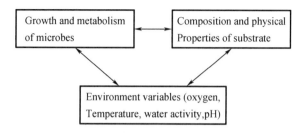

Fig. 5.2 Relationship of the three variables: microbial metabolism, physical properties of substrate, and fermentation parameters

Unit 6

··· Part A ···

Immobilization of Enzymes and Cells

1 Immobilization of Enzymes

Immobilized enzymes have been widely used for synthesis and bioanalysis, even at the industrial level. At this stage, it may be pertinent to critically examine why enzymes in the immobilized form have been so popular. In fact, enzymes, when immobilized, invariably show a decrease in their biological activity. This is due to two factors:

(A) Binding procedure: The various ways by which immobilized enzymes are created may be listed as follows:

(i) Absorptive or ionic binding to carriers or matrices.

(ii) **Covalent coupling** to prefabricated matrices.

(iii) Inclusion methods such as **entrapment** and **microencapsulation**.

(iv) Bioaffinity immobilization.

(v) Chemical aggregates and crosslinked enzyme crystals.

In principle, adsorption and bioaffinity immobilization are more gentle procedures as compared to others. With the advent of fusion proteins, we may see more frequent applications of bioaffinity immobilization. Also, the intended application plays a role in deciding the choice of the binding method. In the case of chemicalcoupling (and, to some extent, in the case of entrapment), the enzyme may **be subjected to harsh** chemicals/extreme pH conditions. In any event, chemical coupling modifies surface residues. These chemical modifications, plus changes in

Unit 6

conformation as a result of **immobilization**, contribute to lower activity after immobilization.

(B) Mass transfer effects: These consist of:

(i) Decreased availability of enzyme molecules within the pores of the carrier.

(ii) **Steric hindrance** by the carrier even if enzyme molecule is on the surface. Slow diffusion by the substrate contributes to these effects.

On the other hand, there are two compensating factors which confer advantages on immobilized forms.

(C) Improved stability: Very often, immobilization results in limited enhancement in stabilization. Sometimes, the observed enhancement of stability may be an artifact especially when **porous** matrices are used. In the beginning, at limited substrate concentration, the more accessible enzyme molecules on the surface are used. As **deactivation** begins, these enzyme molecules cannot use up the substrate which **diffuses in**. The substrate then makes use of enzyme molecules immobilized in the **inner** surface of pores. The latter thus act as a "reserve of fresh activity". The total loss of activity takes more time, giving the **impression** of enhanced stabilization. "In the absence of diffusional restrictions, activity decays exponentially with time, whereas, when diffusional limitations are present, activity decays **linearly** with time."

(D) Reusability: This is, in fact, a far more important factor both in terms of convenience and cost saving. Leaving aside the potential advantage of easier removal of the enzyme from the product, immobilized enzymes so far provide no cost benefit. However, cost savings will be achieved by the repeated reuse of the immobilized enzyme.

New Words

immobilized enzyme	固定化酶
immobilized cell	固定化细胞
covalent coupling	共价交联,共价耦联
entrapment	包埋
microencapsulation	微胶囊,微胶囊技术
be subjected to	受到……,经受……,遭遇……
harsh	刺激性的,严厉的,严酷的,刺耳的
conformation	构象,构造,结构
immobilization	固定,固定化
steric hindrance	位阻,位阻现象,空间障碍

porous	多孔的;可渗透的;有气孔的
deactivation	钝化作用,灭活作用
diffuse in	扩散
inner	内部的,内心的
impression	效果,影响;压痕,印记;感想
decays	衰退,衰减;腐烂,腐朽
exponentially	以指数方式,指数地
linearly	成直线地,线性地
reusability	可重用性,可再使用性

❷ Conjugation Between Enzymes and Smart Polymers

The two main approaches that have been used to create these **hybrid biocatalysts** are adsorption or covalent coupling of the enzyme to the polymer (Tab. 6.1). As in the case of **insoluble** matrices, adsorption, if it works, is a **convenient** and economical way of immobilizing enzymes. Another major advantage is that it does not involve harsh and **toxic** chemicals, which are often required for covalent coupling methods. The trouble is that, even when significant **adsorption** is possible, adsorbed enzymes often come off the matrix in small amounts. **Amylase** adsorbed to **eudragit** did not leak from the matrix; however, repeated uses led to a continuous loss of enzyme activity. "As amylase is sensitive to acidic pH conditions, the decrease in activity observed was most likely due to the **denaturation** of the enzyme when low pH was used for **precipitation**". Thus, multi-covalent linkages between the matrix and the enzyme are necessary to enhance the stability of the enzyme. It is noteworthy that, in the above case, the workers solved the problem by using the covalent coupling method.

In most of the cases, not much effort has been made to optimize **conjugation** conditions. A few guidelines to enhance the efficiency of **carbodiimide** coupling need to be kept in mind: The literature suggests that the coupling proceeds quite well at any pH between 4.5 and 7.5 but that **buffers** contain free **amines**, **sulfhydryls** or **carboxyl** groups should be avoided. **Acetate** and **phosphate** buffers also reduce the reactivity of carbodiimides. Addition of N-**hydroxy sulfo succinimide** is also recommended to enhance coupling efficiency. These aspects have seldom been taken into consideration. For example, phosphate buffer has been employed for carbodiimide coupling with eudragit. It is obvious that more serious efforts are required while developing strategies for obtaining these smart biocatalysts.

Unit 6

Tab. 6.1 Covalent vs. noncovalent binding

Some of the advantages associated with covalent coupling
—Sometimes it is possible to attach more protein covalently than by adsorption
—Oriented immobilization or site-specific conjugation is possible only with covalent coupling
—As a variety of crosslinkers are available, one can almost always obtain an immobilized product with a reasonable level of activity. Adsorption is less predictable, it is not always possible to find a suitable matrix which adsorbs a specific protein
—One can minimize nonspecific binding and covalently link only the desired protein; this is rather difficult in the case of adsorption

New Words

smart polymer	智能聚合物,智能高分子,功能高分子
hybrid	混合的,杂种的
biocatalyst	生物催化剂,酶
insoluble	不能溶解的,不溶性的
convenient	方便的,适当的
toxic	有毒的;中毒的
adsorption	吸附(作用)
amylase	淀粉酶
eudragit	丙烯酸树脂
denaturation	变性,变性作用
precipitation	沉淀,析出,沉淀物
conjugation	结合,共轭
carbodiimide	碳化二亚胺,氨基腈
buffer	缓冲,缓冲液
amine	胺类;有机胺类
sulfhydryl	巯基(等于 sulphydryl)
carboxyl	羧基
acetate	醋酸,乙酸,醋酸盐
phosphate	磷酸,磷酸盐

N-hydroxy sulfo succinimide　　N-羟基硫代琥珀酰亚胺

3 Immobilized Cell Technologies for the Dairy Industry

Traditionally, milk fermentations are conducted in batch bioreactors using freely suspended microbial cells. Lactic acid bacteria (LAB) are widely used in the production of fermented dairy products such as cheeses or fermented milks and creams because of their technological, nutritional and eventual health properties. The production of organic (mainly lactic and acetic) acids and the resulting **acidification** is essential for the production, development of typical flavour and preservation of these products. Other inhibitory compounds such as **bacteriocins** can also increase shelf-life and safety of the products. The transformation of **lactose** by lactic cultures improves the digestibility and various metabolic and enzymatic activities of LAB lead to the production of volatile substances, which contribute to flavour, **aroma** and texture developments in fermented dairy products. **Probiotics** are defined as microbial cells which transit the **gastrointestinal tract** and which, in doing so, benefit the health of the consumer. Among these micro-organisms, **lactobacilli** and **bifidobacteria** are already used in many probiotic dairy products including milk, yogurt, ice cream, and cheese.

There are numerous publications on immobilization of LAB, emphasizing the importance and interest in this new technology. Cell immobilization has been shown to offer many advantages for biomass and metabolite productions compared with free-cell (FC) systems such as: high cell density, reuse of biocatalysts, retention of plasmid bearing cells, improved resistance to contamination, stimulation of production and secretion of secondary metabolites and physical and chemical protection of the cells.

Many advantages have been demonstrated for Immobilized Cells (IC) systems that may be applied to LAB and probiotic bacteria in the dairy and **starter** industries. Application of this research could be particularly important for the production of probiotic bacteria, functional dairy products containing high concentrations of viable bacteria and bio-ingredients from LAB with important functional properties for use in foods and health. Immobilization can efficiently protect cells, making this approach potentially useful for delivery of viable bacteria to the gastrointestinal tract of humans via dairy fermented products. It may be anticipated that application of IC technology in the dairy sector will begin with these special cultures which are difficult to propagate and use with the traditional culture techniques, and which are used to produce highvalue dairy products with positive effects on consumers' health.

Unit 6

New Words

acidification	酸化,使……发酸
bacteriocin	抑菌素
lactose	乳糖
aroma	芳香,香味,香气
probiotics	益生菌,原生菌
gastrointestinal tract	胃肠道;胃肠管
lactobacilli	乳酸杆菌
bifidobacteria	双歧杆菌
starter	发酵剂

4 Organic Acids

The use of immobilized cells has been proposed for a variety of organic acid metabolites relevant to the food industry, including acetic, citric, **fumaric**, lactic, **malic**, **gluconic**, **kojic** and **propionic** acid. In 1994, Norton and Vuillemand reviewed recent progress in organic acid production using immobilized cells. Mori has published an in-depth review of acetic acid production using immobilized cells and, more recently, provided a summary table of bioreactors and cell carriers used for acetic acid production.

The production of citric, malic, tartaric and gluconic acid by fermentation for food applications was reviewed in detail by Milsom. Citric acid production was studied by Sakurai et al. using *Aspergillus niger* cells immobilized on **cellulose beads**. The effects of pre-culture period, initial sugar concentration, and the time interval for repeated batch fermentation cycles on citric acid production were studied. It was found that citric acid production rates and yields were strongly affected by the pre-culture time and conditions. An initial sucrose concentration of 100 g dm-3 was best from the perspective of citric acid production rate and yield. A slight effect of repeated batch cycle time interval, however, an 8-day interval was deemed to be most cost-effective because it had the lowest amount of residual sugar.

Lactic acid is mainly used in the food industry as an **acidulant** and preservative because of its mild acidic taste that does not dominate other flavours in foods. Both single stage and multi-stage reactor configurations have been used for the continuous production of lactic acid with varying degrees of success in terms of lactic acid productivity. As previously mentioned, cell immobilization increases resistance to physicochemical stress, expands the scale of fermentations. In a full-scale lactic acid production plant, a potential stress factor is the presence of **sanitizer** residue in tanks and **piping** that **taint** the bacteria culture medium and reduce activity. Trauth et al. investigated the inhibitory effect of **quaternary ammonium** sanitizers (QAS) on the fermentation activity of lactic acid bacteria Lactococcus lactis. It was found that by immobilizing the lactic acid bacteria in calcium alginate gel beads the inhibitory effects of QAS on cell growth and acidification rate was reduced. As the degree of cell colonization of the beads and the number of successive acidification increased, the acidification rate and resistance to the QAS also increased.

New Words

fumaric	反丁烯二酸的
malic	苹果的,由苹果取得的
gluconic	葡萄糖的
kojic	曲酸
propionic	丙酸的
cellulose beads	球型纤维素
acidulant	酸化剂
sanitizer	食品防腐剂,消毒杀菌剂
piping	管道系统
taint	污染;腐蚀;使感染
quaternary ammonium	季铵;四季铵

5 Immobilized Cell Bioreactor Types for Ethanol Production

The main objectives of immobilization are to increase the bioreactor productivity with improved cell stability, better substrate utilization and easier **downstream processing**. Immobilization also allows for continuous operation and alternative reactor **configurations**. Unless the volumetric productivity of the bioreactor (amount of product formed per unit volume of

Unit 6

bioreactor and time) is strongly increased, the added cost of some immobilization techniques and in some cases higher complexity versus free culture will not be justified, especially for large-scale, low added value products. Thus, associated with cell immobilization is the bioreactor configuration and a good choice of the reactor system is essential for a successful fermentation.

The choice of the bioreactor is related to the type of immobilization, to the metabolism of cells, and the **mass** and heat transfer requirements. For example, the resistance of the matrix to **shear stress**, the size of the beads that affects the mass transfer properties, or the oxygen transfer requirements of the cells may determine the type of reactor. The common immobilized cell bioreactor types used to date for ethanol production are given in Fig. 6.1 – Fig. 6.4.

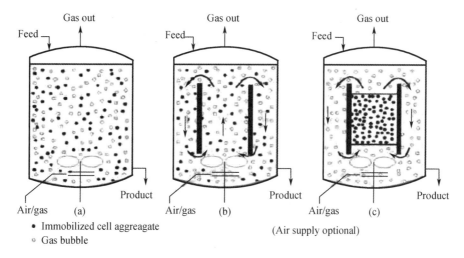

Fig. 6.1 Stirred tank reactors
(a) Simple tank reactor; (b) Draft tube reactor, k; (c) Packed draft tube tank reactor

Although for each type of immobilized cell system a variety of reactor types can be selected, optimal performance requires a careful **matching** of immobilization method and bioreactor configuration. Design of the **cell aggregate** and selection of conditions in the reactor should also go hand in hand.

An added advantage to these immobilized cell bioreactors is that they give much faster fermentation times compared to the existing free cell fermentation. The proven good features of immobilized cell bioreactor systems have been applied to many applications in food and beverages production.

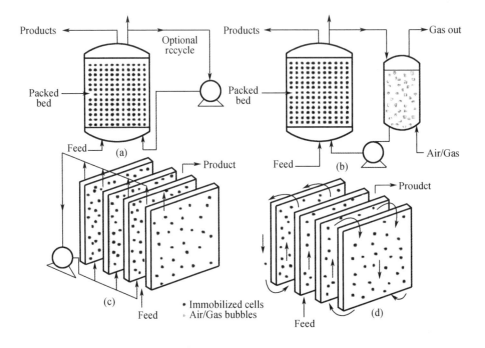

Fig. 6.2 Packed and sheet reactors

(a) Packed bed;
(b) Packed bed with external aeration;
(c) Sheet reactor with external circulation;
(d) Sheet reactor with internal circulation

Unit 6

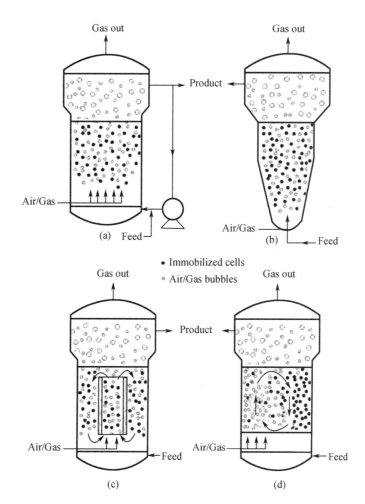

Fig. 6.3 Fluidized bed reactors
(a) Without draft tube; (b) Tapered;
(c) With draft tube; (d) Circulating bed

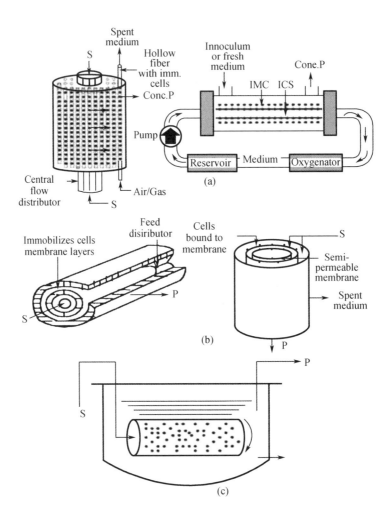

Fig. 6.4 Packed and sheet reactors
(a) Packed bed; (b) Packed bed with external aeration;
(c) Sheet reactor with external circulation; (d) Sheet reactor with internal circulation

Unit 6

New Words

downstream processing	下游处理,下游加工过程,下游技术
configuration	配置
mass	质量
shear stress	剪应力,剪切应力
matrix	基质
match	使相配
cell aggregate	细胞团块,细菌团块

6 Application of Immobilized Cells for Air Pollution Control

The use of pollutant-degrading organisms for air pollution control is an important and emerging application of cell immobilization technology. The principle is relatively simple: a contaminated air stream is passed through a packed bed on which pollutant degrading organisms are immobilized. Contaminants in the air are transferred to the microorganisms, and are degraded to harmless compounds. Air bio-treatment is not a new concept, it has been proposed more than 40 years ago. However, it is only in the past two decades that new environmental regulations have forced engineers to consider alternatives to convention air pollution control methods.

The most successful applications of biological techniques for air pollution control have been for the treatment of dilute, high flow waste gas streams containing **odours** or volatile organic compounds (VOCs). Under optimum conditions, the contaminants are completely degraded to **innocuous end-products**. The major advantage over **conventional** treatment technologies is that air bio-treatment is accomplished at low temperature, and has lower operating and **maintenance** costs.

Biodegradation of the contaminants in bioreactors for air pollution control is usually mediated by mixed cultures or consortia. The primary pollutant-degraders are similar to the organisms found in wastewater treatment processes. They are thriving in a complex and stressful environment that include higher organisms, such as **protozoa**, **rotifers**, even **larvae**, **worms**, **insects** and other predators. The nature of the primary degraders and the fate of the pollutant depend on the main pollutant(s) being treated.

The two most promising bioreactors for air pollution control are bio-filters and **bio-trickling filters** (Fig. 6.5 – Fig. 6.6). In bio-filters, a humid stream of contaminated air is passed through a porous packed bed, usually made of a mixture of **compost** and wood chips or any other bulking agent. On the packing, pollutant-degrading organisms form a bio-film and degrade the absorbed contaminants. Bio-filters are very dry systems with no or little water trickling, hence any metabolite formed during biodegradation will stay in the damp material of the packing. Over time, this may cause inhibition of the process culture. Flushing of the bed is usually effective in removing accumulated metabolites, however it also leaches nutrients and often results in the compaction of the packed bed structure, hence, flushing should be exercised with caution. Bio-filters are simple and cost effective. They require only low maintenance and are particularly effective for the treatment of odour and volatile compounds that are easy to biodegrade, and for compounds that do not generate acidic by-products. Bio-filters are widely used in industrial applications for either VOC or odour control.

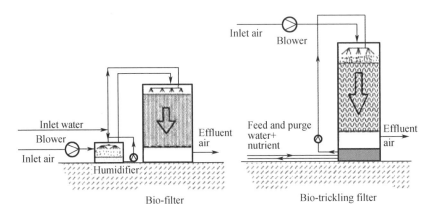

Fig. 6.5 Schematic of bio-filter and bio-trickling filter setups. In-vessel systems are shown, but open bed design is common for bio-filters. The air can be upflow or downflow. The bio-filter shown includes sprinklers for additional moisture supply

Bio-trickling filters work in a similar manner to bio-filters, except that an **aqueous** phase is **trickled** over the packed bed, and that the packing is usually made of a synthetic or inert material, such as plastic rings, open pore foam, lava rock, etc. The trickling solution contains essential inorganic nutrients such as nitrogen, phosphorous, **potassium**, etc. and is either slowly trickled, or trickled at a higher rate and partly recycled. Bio-trickling filters are more complex to build than bio-filters but are usually more effective. They are specially well suited for the

Unit 6

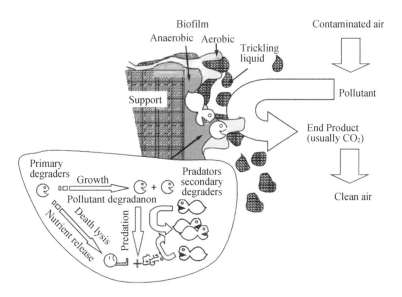

Fig. 6. 6 Simplified treatment mechanism in a bio-trickling filter. Bio-filtration mechanism is similar except that there is no free liquid trickling. Note that in both case, direct gas-bio-film contact exists

treatment of compounds that generate acidic by-products, such as H_2S or **methylene chloride**, because the free aqueous phase allows for a tight control of the conditions such as pH or ionic strength. Another advantage of bio-trickling filters is that they can be built taller than bio-filters because the packing used for bio-trickling filters is usually not subject to compaction. This reduces the required footprint. Bio-trickling filters are more recent than bio-filters, and have not yet been deployed for industrial applications to the same extent as bio-filters.

New Words

odours	气味
innocuous	无害的,无毒的
end-product	最终产品,制成品
conventional	符合习俗的,传统的,常见的,惯例的
maintenance	维护,维修,保养
protozoa	原生动物,原生动物类(protozoan 的复数)

rotifer	轮虫
larvae	幼虫,幼体(larva 的复数形式)
worm	虫,蠕虫
insect	昆虫
bio-filter	生物过滤器,细菌过滤器,生物滤池
bio-trickling filter	生物滴滤池,生物滴滤塔
compost	堆肥,混合物
trickle	滴,细细地流
phosphorous	磷的,含磷的
aqueous	水的,含水的,水般的
potassium	钾
methylene chloride	二氯甲烷,亚甲基氯

Unit 6

··· Part B ···

Applications of Immobilized Enzymes and Cells

1 Starch Hydrolysis

According to a 1998 survey, starch-processing enzymes constitute a global market of US $500 million. This excludes amylases sold for detergents and the textile industries. The major market for starch hydrolysates is in the food industry where they are used as sweeteners andsyrups. They are also fermented further to obtain ethanol, acetone, butanol and lactic acid. Enzymatic hydrolysis of starch to glucose requires α-amylase and glucoamylase, whereas production of industrial sweeteners, high fructose corn syrup, requires glucose isomerase as an additional enzyme (Fig. 6.7). Hoshino et al. immobilized a commercial preparation of amylase, Diabase K-27, to tine types of enteric coating polymers. Best results were obtained with Eudragit L-100. The bioconjugate was completely soluble above pH 5.0. At pH 3.5, the bioconjugate precipitated completely. The specific activity of the immobilized enzyme was 85% of that of the native enzyme. Three approaches were used for starch hydrolysis with this biocatalyst. As these represent generic situations in such systems, it is worthwhile describing them here.

Hoshino et al. have also used amylase immobilized to the smart polymer along with immobilized cells ofLactobacillus casei for continuous lactic acid production from raw starch. The same amylase was linked to hydroxypropyl methylcellulose acetate succinate by the carbodiimide coupling method and Lactobacillus casei was entrapped in kappa-carregeenan. This smart polymer dissolves at pH 5.5 and precipitates at pH 4.0. By optimizing the immobilization conditions, the specific activity of the enzyme was found to be 1.4 times the "best preparation" (obtained with Eudragit earlier).

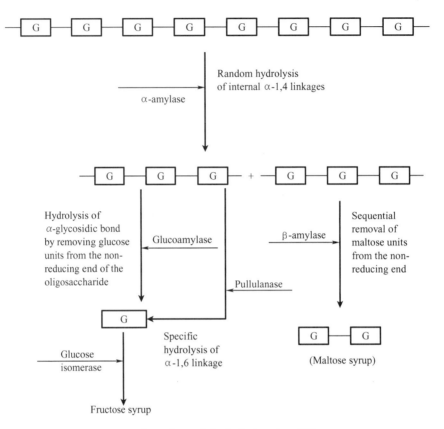

Fig. 6.7 Mechanism of starch hydrolysis using different enzymes

Depending on the mixture of enzymes used, different products of industrial importance may be obtained

2 Hydrolysis of Cellulose

Cellulose, the most abundant organic compound on earth, constitutes a large potential source of feed stock for the fermentative production of fuel alcohol. The intrachain hydrogen bonds result in long and large aggregates called microfibrils. These microfibrils enmesh with lignin and hemicellulose to constitute lignocellulosic material which constitutes over 90% of the dry weight of a plant cell. Enzymatically, cellulosic hydrolysis is carried out by three enzymes. Endoglucanase hydrolyzes some internal bonds in more accessible regions of the microfibrils. Next, cellobiohydrolases attack to generate nonreducing ends to release cellobiose. The disaccharide is hydrolyzed by β-glucosidase to yield glucose.

Taniguchi et al. immobilized a commercial preparation of cellulase fromTrichoderma viride on the methacrylate polymer, Eudragit L. The bioconjugate was completely soluble above pH 5 and insoluble at pH 4. The specific activities of the immobilized enzyme towards microcrystalline cellulose, carboxymethyl cellulose and cellobiose were 63%, 53% and 63% of the free enzyme, respectively. It is believed that as hydrolysis proceeds, cellulose molecules diffuse into the pores of microcrystalline cellulose and become inactivated. It was found that the conjugate of Eudragit-cellulase adsorbed less on the substrate as compared to the free enzyme. This explained the higher activity of the conjugate in the later stages of the reaction. Thus, the size increase upon conjugation, in this case, fortuitously gave a better catalyst. The recycling of the conjugate biocatalyst was carried out in various ways, as described in the case of starch hydrolysis. The most efficient hydrolysis rate was observed when the biocatalyst and unutilized substrate were separated from the product by precipitation, and hydrolysis was recommenced by redissolving the coprecipitate of the biocatalyst and the substrate by adding fresh buffer.

The next challenge in this area will be to use such smart biocatalysts to generate alcohol from cellulose biomass. Among many others that may crop up, two obvious complications will be the inhibitory concentrations of ethanol and control of pH. The latter, of course, could be taken care of by choosing a smart polymer that responds to a stimulus other than pH.

③ Bioaffinity Immobilization

In bioaffinity immobilization, the protein/enzyme is linked to a matrix via biospecific interactions. There are two strategies that are followed for bioaffinity immobilization, which are illustrated in Fig. 5.10. In either case, an affinity pair is involved. In principle, all affinity pairs (affinity ligand-target protein) used in affinity chromatography can also be used for affinity immobilization. However, in affinity chromatography the design of the protocol is geared toward ensuring adequate dissociation (of the target protein), whereas in affinity immobilization the immobilized molecule does not come off the matrix. This adherence can be achieved by choosing affinity pairs that have relatively higher association constants, and/or operational conditions favoring zero leaching. Mattiasson wrote an excellent introduction to the method and listed association constants of some frequently used affinity pairs. More recently, Saleemuddin has reviewed this approach with a focus on affinity immobilization on monoclonal/polyclonal antibodies and lectins. A review with a sharper focus on the use of concanavalin A (Con A)-based supports is also available. It should be mentioned that the classical view of the affinity of biological molecules has now been replaced with a broader concept, according to which a ligand may not have any biological relationship with the target protein, either in vivo or in vitro. Thus, a dye, a metal ion complex, or a peptide may all be useful affinity ligands. In many cases, this provides a successful alternative to more costly affinity ligands such as lectins and monoclonals, although the availability of efficient bioseparation protocols should result in lowering the cost of affinity ligands like lectins and antibodies.

The precoupling of affinity ligand to a matrix can be carried out with any covalent coupling method that is generally used for obtaining affinity media. If the affinity ligand itself is a biologically active protein (e.g., an antibody or a lectin), the design of this precoupling step has to be such that optimum accessibility of binding sites on this affinity ligand (in the precoupled form) to the enzyme (to be immobilized) is possible. Thus, oriented immobilization of proteins as affinity ligands is aimed at linking such affinity ligands (to the matrix) via a site that leaves the binding site free. For example, linking antibodies via F_c portion leaves their binding sites free for interacting with enzymes. A recent work uses site-specific attachment by use of site-directed mutagenesis as another approach to attach the protein affinity ligand in such a way that its binding site is free for "affinity recognition" of the enzyme. Biotin-avidin or biotin-streptavidin technology is now well developed. If the matrix is labeled with avidin/streptavidin or biotin, the enzyme tagged with biotin or avidin/streptavidin, respectively, would allow bioaffinity immobilization with high binding constants. The linking of the enzyme with one of the members of the affinity pair can also be done by chemical cross-linking. Alternatively, fusion proteins can be obtained by recombinant methods.

"Affinity layering" is a relatively recent approach in bioaffinity immobilization. This interesting approach allows large amounts of glycoenzymes to be immobilized by creating alternate layers of Con A and the glycoenzyme.

It is obvious that several innovative variants of bioaffinity immobilization are available. Hence, today, bioaffinity immobilization is a viable and good choice as far as immobilization strategies are concerned.

4 Immobilized Enzymes for Biomedical Applications

The study of immobilized enzymes for biomedical applications started in the 1960s, aiming to solve some of the limitations to the use of enzymes in clinics, to make them more stable, less

Unit 6

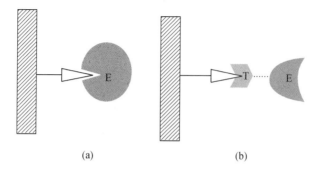

Fig. 6.8 Two commonly used strategies for bioaffinity immobilization of proteins
(a) The enzyme (E) has affinity for an affinity ligand (-β) linked to the matrix; (b) The enzyme (E) is linked to a fusion tag (T), which has affinity for an affinity ligand (-β) linked to the matrix

immunogenic and toxicologic, and to present a longer in vivo circulation lifetime. Since then, several approaches have been used in enzyme therapy either for the detection of bioactive substances in the diagnosis of diseases or with the aim to treat a disease condition, such as the correction of inborn metabolic defects, cardiovascular diseases, cancer, intestinal diseases, or for the treatment of intoxication.

For the immobilization of enzymes two different approaches have been used, the first consisting in the cross-linking or the covalent attachment of the enzyme to a support (immobilization by binding) and the second based on the entrapment of the enzyme in a matrix (immobilization by inclusion).

Among the methods used for immobilization by binding, conjugation with polymers has received great attention in the last several years. A large number of polymers can be considered for enzyme immobilization, although the strict requirements needed for a biomedical approach limit this number significantly. The polymers used for enzyme immobilization may be fully biocompatible, and biodegradable, and should possess high purity and homogeneous molecular-weight (MW) distribution. Besides, one should bear in mind that covalent coupling of enzymes to polymers may result in conformational alterations and decrease significantly enzymatic activity.

The immobilization by inclusion methods present some advantages over the methods used for the immobilization by binding: there is no need for derivatization of the enzyme, the systems provide a higher degree of protection against enzymatic degradation and other destructive factors, allow for much higher drug loads, and enzymes can be immobilized into the system in a more stabilized form, such as multi-enzymatic systems, enzyme mixtures, or even cells producing a certain enzyme.

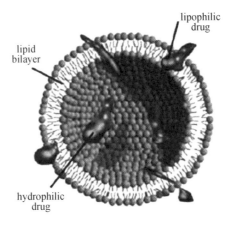

Fig. 6.9 Structure of a liposome. Hydrophilic drugs are encapsulated in the inner aqueous space, whereas lipophilic drugs are entrapped in the phospholipid membrane

However, inclusion methods present disadvantages related to the use of synthetic materials, which can be solved by the use of biodegradable polymers (e. g. , polymers and copolymers derived from lactic and glycolic acid, alginate, chitosan) or by the use of more biocompatible immobilization carriers such as liposomes or red blood cells.

Among particulate drug carriers, liposomes are the most extensively studied and possess the most suitable characteristics for polypeptide encapsulation. Liposomes are microscopic vesicles composed of membrane-like phospholipid bilayers enclosing aqueous compartments (Fig. 6.10). They are biologically inert, biocompatible, and cause very little toxic or antigenic reaction. By the encapsulation of enzymes into liposomes, the liposomes can mask their antigen determinants, avoiding the adverse immunological reactions.

5 Immobilization of Enzymes for Use in Organic Media

As a result of intense research over the past two decades, it is now generally realized that many enzymes express catalytic activity in organic media, and many useful applications in synthesis have therefore been presented. "organic media" is defined as media consisting predominantly of organic solvents or other organic substances. Solvent free media, in which the

Unit 6

substrates to be converted function as solvents as well, constitute an important example. The organic media contain small amounts of water, which the enzymes need in order to express catalytic activity. Ionic liquids and supercritical fluids have also been used as media for enzymatic reactions. In these applications, they share several characteristics with organic media, mainly because of the low water content in all these media.

Proteins are soluble in very few organic solvents. This means that in the majority of the organic media the enzymes are insoluble. It is therefore unnecessary to immobilize the enzymes in order to make it easier to separate them from the reaction mixture after the reaction. In many small-scale reactions, enzyme powders have been used as catalysts. However, immobilization is usually beneficial in order to improve the properties of the enzyme preparation. It is therefore quite common to immobilize enzymes by adsorption or deposition on porous supports before using them in organic media. The support material can be chosen so that it has suitable properties to be used in the intended reactor. In mechanically stirred reactors the materials should be resistant to shear forces, whereas in packed bed reactors the compressibility is more important because of its influence on the pressure drop of the reactor. In general, immobilization of the enzyme on a suitable support increases the observed catalytic activity as a result of effective dispersion of the enzyme on the surface of the support.

Enzyme immobilization methods for organic media are generally quite simple. An aqueous solution of the enzyme is mixed with the support, and the water is removed so that the enzyme is deposited on the surface of the support. The pH value of this solution is important. It has been said the enzyme "remembers" the pH of the last aqueous solution in which it was dissolved. The ionization state of the enzyme will remain essentially unchanged-unless acids or bases are present in the reactant solution-once the water has been removed and the enzyme preparation is put into an organic medium. Usually the aqueous solution used for enzyme immobilization is thus buffered to a pH value close to the pH optimum of the enzyme.

In some cases, enzymes adsorb spontaneously on the support. For example, it is possible to adsorb lipases efficiently on hydrophobic supports, such as porous polypropene. In this case, it is practical to use a rather large volume of aqueous solution in contact with the support. Lipases and other hydrophobic substances adsorb on the support whereas other substances remain in the solution. After complete adsorption, the immobilized enzyme is recovered by filtration. In this method, polar impurities in the enzyme powder used to prepare the solution are efficiently removed and thus immobilization simultaneously constitutes an enzyme purification step.

Not all enzymes are suitable for immobilization by adsorption. Some enzymes are adsorbed too strongly and thereby lose activity whereas others are not adsorbed efficiently enough. For the latter case, "deposition" constitutes a useful alternative method. Here, the enzyme is dissolved in a smaller volume of aqueous buffer, which is mixed with the support and followed by drying of the complete mixture. In this case, everything present in the solution is deposited on the support. This procedure has been found useful for a wide range of enzymes on a wide range of supports. Celite is a typical support used in the deposition method.

6 Immobilization of Lipases

Lipases have as natural function the hydrolysis of triglycerides. Nevertheless, they may be used in vitro to catalyze many different reactions, in many instances quite far from the natural ones (e.g., regarding conditions, substrates). Thus, lipases are utilized in different industrial applications, such as in the production of modified oils, cosmetics, or the most important one in the last years-production of many different intermediates for organic synthesis (e.g., resolution of racemic mixtures), because they combine broad substrate specificity with a high enantio- and regioselectivity.

However, these enzymes present a very complex catalytic mechanism. In homogeneous aqueous solutions, the lipase is mainly in a closed and inactive conformation where the active site is completely isolated from the reaction medium by an oligopeptide chain (called flap or lid), blocking the entry of substrates to the active site. This polypeptide chain presents several hydrophobic amino acid residues in its internal face, interacting with hydrophobic zones around the active site. This conformation may exist in a partial equilibrium with an open and active conformation, where the lid is displaced, stabilized by ionic interactions or hydrogen bonds with a specific part in the lipase surface allowing the access of the substrate to the active site.

However, upon exposure to a hydrophobic interface such as a lipid droplet, the lipase can only interact (when it is in the open conformation), via the hydrophobic pocket formed by the internal face of the lid and the surroundings of the active site, consequently shifting the equilibrium towards the open form-the so called interfacial activation (Fig. 6.10).

This mechanism of action promotes some problems when industrial immobilized lipases are prepared. Inside a porous structure, lipases molecules become inaccessible to any kind of external interfaces, therefore there is no possibility of enzyme interfacial adsorption in aqueous solutions. In fact, conventionally immobilized lipase preparations are usually utilized primarily in anhydrous media, where the lipase may become activated by the direct interaction with the organic solvent phase.

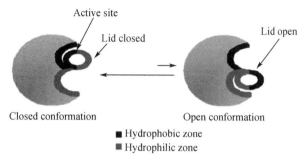

Fig. 6.10　Interfacial adsorption of lipases

However, many interesting reactions catalyzed by lipases may be advantageously carried out in aqueous systems (e. g., hydrolytic resolutions of racemic mixtures). The interfacial adsorption of lipases on hydrophobic supports has been proposed as a simple method for preparing immobilized lipase preparations useful in any media. The hypothesis behind this immobilization strategy is to take advantage of the complex mechanism of lipases (an apparent problem) as a tool that permits the immobilization of lipases via an "affinity-like" strategy. Using a hydrophobic support (that somehow resembles the surface of the drops of the natural substrates) and very low ionic strength, lipases become selectively immobilized on these supports. These adsorbed lipases are able to access the active center; in fact, immobilized enzymes usually exhibit significantly enhanced enzyme activity. The result is an immobilized lipase where the open conformation has been "fixed" and does not depend on the presence of external hydrophobic interfaces.

Unit 7

··· Part A ···

Protein Extraction Using Reverse Micelles

1 Introduction

In order to be exploitable for **extraction** and **purification** of proteins/enzymes, **RMs** should exhibit two characteristic features. First, they should be capable of **solubilizing** proteins **selectively**. This protein uptake is referred to as forward extraction. Second, they should be able to release these proteins into aqueous phase so that a **quantitative** recovery of the purified protein can be obtained, which is referred to as **back extraction**. A **schematic** representation of protein **solubilization** in RMs from aqueous phase is shown in Fig. 7.1. In a number of recent publications, extraction and purification of proteins (both forward and back extraction) has been demonstrated using various reverse micellar systems. In some references, exclusively various enzymes/proteins that are extracted using RMs as well as the stability and **conformational** studies of various enzymes in RMs are summarized. The studies revealed that the extraction process is generally controlled by various factors such as **concentration** and type of **surfactant**, pH and **ionic strength** of the aqueous phase, concentration and type of co-surfactants, salts, charge of the protein, temperature, water **content**, size and shape of reverse micelles, etc. By manipulating these parameters selective separation of the desired protein from mixtures can be achieved. These parameters are discussed below.

Unit 7

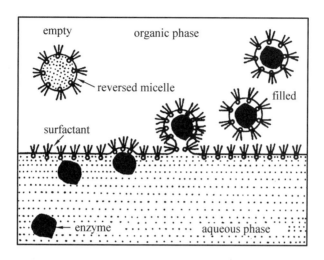

Fig. 7.1 Schematic representation of mechanism of protein solubilization into reverse micellar phase from aqueous phase

New Words

micelle	胶束,胶囊,微胞,微团
extraction	提取(法);萃取法
purification	纯化
RMs (reverse micelles)	反转胶团
solubilize	(使)溶解;(使)增溶
selective	选择的,选择性的
quantitative	数量的,定量的
back extraction	反萃取
schematic	示意性的
solubilization	溶解;增溶(作用)
conformational	构象的
concentration	浓度
surfactant	表面活性剂(的)
ionic strength	离子强度

· 147 ·

content　　　　　　　　　　比例(某种特定物质的比例)

2　Water Content and Water Pool

The water content of the RMs is defined as the ratio of the water molecules to that of the surfactant molecules per RM ($W_0 = [H_2O]/[surfactant]$). W_0 strongly depends on the relative solubility of surfactant in the **polar** and **nonpolar solvents**, expressed as **hydrophilic-lipophilic balance** (**HLB**) of the surfactant and it increases with HLB. The HLB increases in the following manner for cationic surfactants: **TOMAC < DDAB < BDBAC < CTAB ~ CPB**. W_0 has a major role in protein solubilization and function. Krei and Hustedt have observed a significant influence of W_0 or size of the RM on the partitioning behavior of proteins (e.g. α-**amylase**) in various reverse micellar systems containing **cationic surfactants**. They found that below a certain **critical value** of W_0 (~40), the **physicochemical** properties of the aqueous **microphase** strongly depends on the micellar size, thereby favoring the partition **coefficient** of the protein towards the excess aqueous phase. Hilhorst et al. have concluded that W_0 of the TOMAC/**octane** system can be varied by exchanging surfactant **counterions** with ions in the aqueous **bulk solution**, and by changing the amount of alcohol or co-surfactant, or nature of co-surfactant. Further, they found that increasing W_0 (up to a certain value) facilitates transfer of α-amylase into the RMs and transfer is observed at a lower pH. In a recent study, it was reported that a high W_0 (~120) was required to achieve almost complete solubilization of **inulinase** into BDBAC/**isooctane/hexanol** RMs.

The nature of the water in the core of the reverse micelle is of great importance since proteins/enzymes and other **biomaterials** reside in this water pool. The water pool is generally regarded to be a composite of two different types, the bound water (lining the interior wall of the **AOT** micelles) and the (remaining) free water. Further subdivisions of the water pool have also been proposed. Gierasch et al. used IR **spectra**. It should be stressed that water entrapped in RMs is different from bulk water and is similar to water present in the vicinity of biological membranes or proteins in that it has restricted mobility, depressed **freezing point**, and characteristic spectroscopic properties. The unusual behavior of this water has been attributed to its **strong interaction** with the head groups of the surfactant as well as to an overall disruption of the three-dimensional **hydrogen-bonded** network usually present in bulk water.

Unit 7

New Words

polar	极性的
nonpolar	无极性的
solvent	溶剂
hydrophilic	亲水的,吸水的
lipophilic	亲脂性的
hydrophilic-lipophilic balance (HLB)	亲水－亲脂平衡
TOMAC (trioctylmethyl ammonium chloride)	三辛基甲基氯化铵
DDAB (didodecyldimethyl ammonium bromide)	双十二烷基,甲基溴化铵
BDBAC (N-benzyl-N-dodecyl-N-bis (2-hydroxy ethyl) ammonium chloride)	氮－苯基,十二烷基,双（2－羟乙基）氯化铵
CTAB (Cetyltrimethyl Ammonium Bromide)	溴化十六烷三甲基铵
CPB (cetyl pyridinium bromide)	十六烷基吡啶溴酸盐
amylase	淀粉酶
cationic surfactant	阳离子表面活性剂
critical value	临界值
physicochemical	物理化学的
microphase	微相
coefficient	系数
octane	辛烷
counterion	带相反电荷的离子；抗衡离子,补偿离子
bulk solution	本体溶液
inulinase	菊粉酶
isooctane	异辛烷
hexanol	己醇
biomaterial	生物材料
AOT (sodium bis(2-ethyl-1-hexyl) sulfosuccinate)	双(2－乙基－1－己基)硫代丁二酸钠
spectra	光谱
freezing point	凝固点,凝固温度
strong interaction	强相互作用

hydrogen-bonded   氢键

3 Aqueous Phase pH

The aqueous phase pH determines the **ionization** state of the surface-charged groups on the protein molecule. Solubilization of the protein in RMs is found to be dominated by **electrostatic interactions** between the charged protein and the inner layer of the surfactant head groups. Solubilization of protein is favored at pH values above the **isoelectric point** (**pI**) of the protein in the case of cationic surfactants, while the opposite is true for anionic surfactants. Chang et al. showed that the solubilization of α-amylase in **aliquat**-336 could be achieved by increasing the pH above its pI (5.4). They observed a maximum solubilization (~85%) of the enzyme at pH 10. Many other studies also indicated dependence of protein solubility on pI. It may be noted that for proteins with small molecular weight such as **cytochrome** C, **lysozyme**, and **ribonuclease** (MW range 12,000 – 14,500 Da), the (pH – pI) value required for optimum solubilization is much lower (<2) when compared to that of larger proteins such as α-amylase (MW 48,000 Da) and **alkaline protease** (MW 33,000 Da) where (pH – pI) is around 5. This can be explained with the reasoning that as the protein size increases, size of the RM also has to increase in order to **incorporate** the protein molecule. To increase the size of RM, higher energy is required which can be provided by increasing the number of charged groups on the protein. This increase in charge **density** on the protein molecule can be accomplished by manipulating the pH of the aqueous solution (i.e., by increasing the pH much higher than the pI of the protein). For small proteins whose size is smaller than the size of the water pool inside an RM, solubilization occurs as soon as the net charge is opposite to that of the reverse micellar interface.

New Words

ionization	离子化,电离
electrostatic interactions	静电相互作用
isoelectric point (pI)	等电点
aliquat	季铵氯化物
cytochrome	细胞色素

lysozyme	溶解酵素
ribonuclease	核糖核酸酶
alkaline	碱的,碱性的
protease	蛋白酶
incorporate	混合
density	密度

4 Ionic Strength

The influence of **ionic strength** (KCl/NaCl concentration) on the solubilization of proteins in RMs is explained purely as an electrostatic effect. The electrostatic potential of a protein molecule in an **electrolyte** is inversely **proportional** to the ionic strength of the solution and is characterized by **Debye** length. In general, it was observed that as the ionic strength of the aqueous solution increases, the protein intake capacity of the RMs decreases. Two reasons were given to explain this phenomenon. First, increasing the ionic strength decreases the Debye length thereby reducing the electrostatic interaction between the charged protein molecules and charged surfactant head groups of the RMs. Second, increasing the ionic strength reduces the **electrostatic repulsion** between the charged head groups of the surfactants in a RM, thereby decreasing the size of RM. The smaller RMs will have larger **curvature**, which increases the density of the surfactant **monolayer** near the surfactant head groups, resulting in a gradual **expulsion** of the protein molecules residing inside the RMs. The process is termed as a **squeezing-out** effect. Further, it has been observed that not only the concentration but also the type of the ions play a very important role in determining W_0 and partition behavior of proteins in RMs. It may be noted that although lower side of ionic strength favors the protein transfer, one cannot perform the experiments at very low ionic strengths (NaCl/KCl concentration $\leqslant 0.01$ mol L^{-1}) because under these conditions the solution becomes cloudy. Further, Marcozzi et al. have stated that the type and concentration of salt used in the forward extraction process remarkably affect the percentage recovery and activity in the back extraction process.

New Words

ionic strength	离子强度
electrolyte	电解,电解液
proportional	比例的,成比例的
Debye	德拜(偶极矩的单位)
electrostatic repulsion	静电排斥
curvature	弯曲,曲率
monolayer	单层
expulsion	逐出
squeeze-out	挤(出去),冲(出去)

5 Surfactant Type

The protein **distribution** is mainly dependent on the **charge difference** between the protein and the surfactant head groups. When other effects are insignificant, pH of the protein solution determines the distribution behavior of protein in RMs stabilized by charged surfactants. In addition to the charge, other surfactant-dependent parameters such as the size of RMs, the energy required to enlarge the RMs, and the charge density on the inner surface of the RMs may also influence the protein distribution.

Most of the work on RMs reported to date has been with AOT, which is an **anionic** surfactant. Cationic surfactants such as **quaternary ammonium** salts (TOMAC and CTAB) have also been used for protein solubilization. The studies with AOT RMs in many cases have shown a rapid degradation of the protein activity after their solubilization. Protein **denaturation** was also observed in case of RMs having cationic surfactants. For instance, **glucose-6-phosphate dehydrogenase** was denatured in CTAB/hexanol/octane RMs. Recently, many researchers have studied protein transfer using **non-ionic** surfactants such as **Tween**-85. It was demonstrated that a non-ionic surfactant has an apparent advantage over ionic surfactants due to the absence of strong charges at the aqueous/organic interface, which provide a suitable environment for the protein. Further, Tween-85 is **nontoxic**. Although, a few studies have been carried out using non-ionic surfactants and shown to alleviate significantly protein **stability** problems, there has been little understanding regarding the reason for enzyme stability (with respect to structure and function) and solubilization in these RMs. In many cases the effect of temperature, pressure, and ionic strength were found to be opposite to those observed in ionic based mixtures.

New Words

distribution	分配
charge difference	电荷差
anionic	阴离子的,带负电荷的离子的
quaternary ammonium	季铵
denaturation	使变性
glucose phosphate	葡萄糖磷酸
hydrogenase	氢化酶
non-ionic	非离子物质
Tween	吐温
nontoxic	无毒的
stability	稳定性

6 Surfactant Concentration

The concentration of surfactant has been shown to have a little effect on the structure and size or aggregation number of the RMs. However, it causes changes in the number of RMs, which increases the protein solubilization capacity of the RMs. Enhanced amino acid and protein solubilization were observed in RMs by increasing the TOMAC (up to 200 mmol L^{-1}) and BDBAC (up to 150 mmol L^{-1}) concentration respectively. However, further increase in the surfactant concentration decreased solubilization of **biomolecules**. At surfactant concentration above a certain value, the micellar interactions may occur leading to percolation and **interfacial deformation**, with a change in the micellar shape and micellar **clustering**. Goklen suggested that at high surfactant concentrations, **monodisperse** spherical micelles might not be present predominantly in the solution. The micellar clustering decreases the interfacial area available to host the biomolecules causing a decrease in the solubilization capacity of the RMs. This is especially true for solutes with strong interfacial interactions, such as **tryptophan**. Alexandridis et al. and Cardoso et al. have shown that AOT concentration is related to percolation. The percolation phenomenon was followed by a steep increase in the micellar **conductivity**. Huang and Lee observed a drastic reduction in the recovery (around 40%) of the horse radish **peroxidase** when AOT concentration was at 5 mmol L^{-1} and further increase in AOT concentration to 10 mmol L^{-1} produced no recovery at all.

New Words

biomolecule	生物分子
percolation	过滤,浸透
interfacial	界面的;分界面的,面间的
deformation	变形
clustering	聚类
monodisperse	单分散(性)的
tryptophan	色氨酸
conductivity	传导性,传导率
peroxidase	过氧(化)物酶

7 Micelle Size

Micelle size is dependent on the ratio $W_0 = [H_2O]/[Surfactant]$ but not on $[H_2O]$ or [Surfactant] alone. The monodisperse small sized RMs can accommodate only proteins of certain **dimensions**. Hence, micelle size may be used to include or **exclude** certain proteins. However, it should be noted that several micelles can **regroup** to form larger micelles when certain operating conditions are altered. It was also **hypothesized** that a protein can create around itself a new larger micelle of a requisite size to facilitate solubilization. Wolf and Luisi reported that a given protein induces the formation of large enough micelles for the protein to fit into. Besides water and surfactant concentration, it was also demonstrated that **equilibrium** micelle size can be controlled by ionic strength of the **bulk aqueous phase**. As the ionic strength increases the micellar size decreases due to decrease in the **electrostatic repulsion** between the head groups of surfactants. This results in gradual expulsion of the protein molecule from the RMs, which is termed as a

Unit 7

squeezing out effect. Besides ionic strength, micelle size is also influenced by the type of the solvent. Therefore selection of a suitable solvent becomes an important task and several solvents should be tested. Goklen and Hatton studied the effect of solvent structure on maximum RM size, W_0^{max}. **Hexane**, **isooctane** and octane were reported to have much higher W_0^{max} values (75 – 115) compared to **dodecane**, **cyclohexane**, **xylenes**, **carbon tetrachloride**, or **chloroform** (5 – 20).

New Words

dimension	尺寸,尺度
exclude	排斥
regroup	重组,重编
hypothesize	假设
equilibrium	平衡
bulk aqueous phase	水相
electrostatic repulsion	静电排斥
hexane	(正)己烷
isooctane	异辛烷
dodecane	十二烷
cyclohexane	环己胺
xylene	二甲苯
carbon tetrachloride	四氯化碳
chloroform	氯仿

8. Miscellaneous Factors

Other factors that may affect protein extraction are **volume ratio** of organic to aqueous phases (V_{org}/V_{aq}), co-surfactant, temperature, **mass transfer** efficiency, protein charge, electrostatic potential of the RMs, and presence of other ions such as Ca^{2+}, Mg^{2+}, Ba^{2+}, etc..

(V_{org}/V_{aq}) is a **critical parameter** in extraction and concentration of enzymes. For an ideal system, in general, this ratio should be low for extraction and high for stripping steps to achieve concentration. The use of a cosurfactant may enhance the solubilization **kinetics**, stability of a RM, and even selectivity. Medium-chain length alcohols like **isopropanol** and hexanol have been used for this purpose. Temperature is a factor to investigate within the range of enzyme stability and **inactivation**. It was reported that as temperature increases, water solubilization capacity increases in the organic phase in case of AOT-RMs, and the exactly opposite phenomenon occurs in the case of TOMAC-RMs. The variation in the **water uptake** is due to a change in the aggregation number of the surfactants. It was also noted that to maintain a clear reverse micellar phase after forward extraction, temperature control (25 ℃) was critical. In contrast to these observations, Huang and Chang have recently reported that, as temperature increases, AOT content in the organic phase reduces, resulting in a poor forward **transfer efficiency**. They also observed that back extraction of the enzyme α-**chymotrypsin** was favored at higher temperature (40 ℃). Two postulates were made for this enhanced enzyme recovery: (i) the RMs will be disrupted at higher temperatures due to their high energy level, and the relaxation time for enzyme transfer through oil-water interface will be decreased and (ii) the **migration** rate of the AOT-RMs through the oil-water interface toward the water phase will be higher at higher temperatures, which leads to enhanced recovery of the enzyme activity. Protein charge and electrostatic potential within the RMs are closely linked to factors such as pH and ionic strength. In addition to the chemical and biochemical factors, protein transfer also depends on physical aspects such as extent of **mixing** and contact surface between the two phases.

Presence of ions was also found to have an effect on protein solubilization. Misiorowski and Wells studied the solubilization and the activity of **phospholipase** A_2 in RMs and found that **incorporation** of the enzyme into micelle was greatly dependent on the presence of divalent **cations**. Though Ca^{2+} was absolutely necessary for specific catalytic function, solubilization could also be achieved by Mg^{2+} and Ba^{2+}. Regalado et al. have observed a poor solubility of **horseradish** peroxidase in AOT-RMs mainly due to the presence of high mineral content in the crude extract, which has reduced the W_0 value of RMs significantly. The ions caused an **electrostatic screening** of the surfactant head groups hindering their interactions with the protein molecule, leading to reduced W_0 values. Addition of calcium **scavengers** (**citric acid-citrate** and **EDTA-disodium** salt) improved the solubilization and recovery of the enzyme.

New Words

miscellaneous	多方面的
volume ratio	体积比

mass transfer	质量传递[转移],传质
critical parameter	临界参数
kinetics	动力学
isopropanol	异丙醇
inactivation	灭活,失活
water uptake	吸水
transfer efficiency	转换效率
chymotrypsin	胰凝乳蛋白酶,糜蛋白酶
migration	移动
mixing	混合
phospholipase	磷脂酶
incorporation	结合
cation	阳离子
horseradish	辣根
electrostatic screening	静电屏蔽
scavenger	净化剂
citric acid	柠檬酸
citrate	柠檬酸盐
EDTA	乙二胺四乙酸
disodium	二钠盐

 Mechanism and Methods of Protein Solubilization

Although many studies have been performed on the **RME** of proteins and the catalytic properties of enzymes in RMs, very little is known about the mechanism of protein solubilization in RMs and the major **driving forces** which affect the solubilization.

There are three commonly used methods to incorporate enzymes in RMs: (i) **injection** of a concentrated aqueous solution, (ii) addition of dry lyophilized protein to a reverse micellar solution, and (iii) phase transfer between bulk aqueous and surfactant-containing organic phases. Fig. 7.2 shows schematically the three enzyme solubilization methods. The injection and dry addition techniques are commonly used in biocatalytic applications, the latter being well suited to hydrophobic proteins. The phase transfer technique is the basis for extraction of proteins from aqueous solutions.

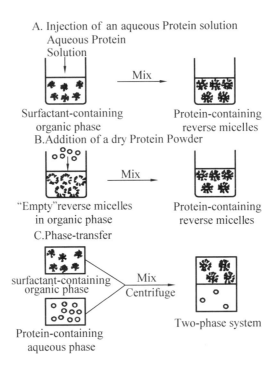

Fig. 7.2 Various methods of protein solubilization in reverse micelles

Several experimental and theoretical studies have been reported on the solubilization of proteins in RMs. Experimental tools used to study protein solubilization include **ultracentrifugation, quasi-elastic light scattering (QELS), small-angle neutron scattering (SANS), small-angle X-ray scattering (SAXS)**, and **fluorescence** recovery after **fringe pattern photobleaching**. Protein structure in reverse micellar phase has been examined with **ultraviolet (UV)** and fluorescence **spectroscopies** as well as optical rotary dispersion and **circular dichroism**. High-pressure **electron paramagnetic resonance spectroscopy** was also used to assess protein mobility in RMs.

Blanch and coworkers investigated in detail the solubilization properties of α-chymotrypsin and **alcohol dehydrogenase (ADH)** in RMs prepared by the above three techniques. Protein solubilization in RMs greatly depends on the method used for protein addition as well as on the size of the protein and of the RM. For the dry addition method protein solubilization is strongly dependent on micelle size whereas for the injection method it is less dependent. For smaller proteins like α-chymotrypsin (**diameter** of 44Å), maximum solubilization occurred when the micelle diameter was 50 – 60Å. For larger proteins like ADH (dimensions of 45 * 60 * 110Å), maximum solubilization occurred at a larger micelle size of 80 – 90Å. Therefore, when a dry protein powder is added, the size of the RM must be approximately the same or larger than the protein molecule for efficient solubilization. It appears that the energy barrier for solubilization of a large protein in a small micelle is too large to overcome. For a larger micelle, since the micelle is not required to rearrange its contents to incorporate a protein, the **energy barrier** is lower and the protein is solubilized. Matzke et al. also revealed that, at intermediate micelle diameters, the reverse micellar system enhances the solubility of ADH above a value that is attainable in bulk aqueous phase. This enhanced solubility suggests that ADH interacts with the reverse micellar interface. Vos et al. reported electrostatic interactions between ADH and AOT interface, which was supported by Matzke et al.

For the injection method, where a **saturated** protein **solution** is added directly to the surfactant-containing organic phase, the RMs are forced to form around the proteins already present in the solution. Thus, protein solubilization is not strongly dependent on micelle size in this technique.

The phase transfer method of protein solubilization is fundamentally different from the other two methods. In this method, there are two **bulk phases** (aqueous and organic) which are brought to equilibrium. Under certain conditions, the protein molecules are transferred from the aqueous phase to the surfactant-containing organic phase. Unlike the dry-addition and injection methods, it is difficult to obtain a value for the maximum solubilization using the phase-transfer method. Moreover, since this method is mainly used for protein extraction, it is desirable to use aqueous phase protein concentrations consistent with those in a typical fermentation broth. For the phase-transfer method, pH of the aqueous phase, the size and isoelectric point of the protein, and the surfactant type were shown to have a significant effect on protein solubilization.

Cardoso et al. using the phase transfer technique studied the driving forces involved in the selective solubilization of three different amino acids having same pI, namely aspartic acid (hydrophilic), phenylalanine (slightly hydrophobic), and tryptophan (hydrophobic) in cationic TOMAC-RMs. The main driving forces involved were found to be hydrophobic and electrostatic interactions. Few other researchers have also identified that the major driving forces involved in the amino acid solubilization were hydrophobic interactions and amino acid structure as well as its **ionization** state.

New Words

RME (reverse micellar extraction)	反相胶束抽提
driving force	驱[传,主]动力
injection	注射,注射剂
ultracentrifugation	超速离心法
quasi-elastic light scattering (QELS)	类似弹性光散射
small-angle neutron scattering (SANS)	小角中子散射
small-angle X-ray scattering (SAXS)	小角X-光散射
fluorescence	荧光,荧光性
fringe pattern	干涉图
photobleaching	光(致)褪色(漂白)
ultraviolet (UV)	紫外线的,紫外的；紫外线辐射
spectroscopy	光谱学,波谱学
circular dichroism	圆形(循环)二色性
electron paramagnetic resonance spectroscopy	顺磁共振光谱学
alcohol dehydrogenase (ADH)	醇脱氢酶
diameter	直径
energy barrier	能障;能垒
saturated solution	饱和溶液
bulk phase	体相
ionization	离子化,电离

10 Back Extraction

Transfer of solubilized proteins from the reverse micellar phase back to an aqueous phase constitutes back extraction. A successful RME should include both forward and back extraction processes in their optimized conditions. In contrast to the extensive studies investigating the forward extraction process, back extraction has been addressed to a much lesser extent. Most of the earlier studies tacitly assume that conditions, which normally prevent protein uptake in the forward

Unit 7

transfer, would promote their release in the back transfer. That is to select a pH and salt condition that had minimal forward transfer efficiency. This assumption, however, is not true and resulted in only a low protein recovery. As reviewed by Kelley et al. , overall recovery in RME is generally below 80%. Rahaman et al. reported only 10% - 20% of alkaline protease recovery from AOT/isooctane RMs, which was attributed to kinetic **limitations**.

Recently, some alternative approaches for enhanced recovery of the proteins from RMs have been investigated. They include: (i) use of **silica** particles for the **sorption** of the proteins as well as surfactants and water directly from the protein-filled RMs and use of ion-exchange columns, (ii) addition of **dewatering** agents such as **isopropyl alcohol** and dehydration of RM with **molecular sieves** to recover the protein, (iii) addition of large amount of a second organic solvent, such as **ethyl acetate** to destabilize the RM and hence to release the protein, (iv) formation of clathrate **hydrates** via **pressurization**, (v) use of temperature to dewater the RMs and hence to release the protein, (vi) use of NaDEHP/isooctane/**brine** RMs which can easily be destabilized by adding **divalent** cations (such as Ca^{2+}) and subsequent release of the protein into aqueous media, (vii) addition of sucrose to enhance the protein recovery by reducing the protein-surfactant interactions, and (viii) back extraction with the aid of a counter-ionic surfactant and through the gas **hydrate** formation.

New Words

back extraction	后萃取；反萃取
limitation	限制
silica	硅石，无水硅酸，硅土
sorption	吸附作用
dewatering	去水(作用)，脱水(作用)
isopropyl alcohol	异丙醇
molecular sieve	分子筛
ethyl acetate	乙酸乙酯
hydrate	氢氧化物；与水化合
pressurization	增压；加压
brine	盐水
divalent	二价的
hydrate	氢氧化物；与水化合

11 Mass Transfer Kinetics

Rate of protein transfer to or from a reverse micellar phase and factors affecting the rate are important for the **practical applications** of RME for the extraction and purification of proteins/enzymes and for **scale-up**. The mechanism of protein exchange between two **immiscible phases** (Fig. 7.1) can be divided into three steps: the **diffusion** of protein from bulk aqueous solution to the interface, the formation of a protein-containing micelle at the interface, and the diffusion of a protein-containing micelle in to the organic phase. The reverse steps are applicable for back transfer with the coalescence of protein-filled RM with the interface to release the protein. The overall mass transfer rate during an extraction processes will depend on which of these steps is rate limiting.

Dekker et al. studied the extraction process of α-amylase in a TOMAC/isooctane reverse micellar system in terms of the **distribution coefficients**, mass transfer coefficient, inactivation rate constants, phase ratio, and residence time during the forward and backward extractions. They derived different equations for the concentration of active enzyme in all phases as a function of time. It was also shown that the inactivation took place predominantly in the first aqueous phase due to complex formation between enzyme and surfactant. In order to minimize the extent of enzyme inactivation, the steady state enzyme concentration should be kept as low as possible in the first aqueous phase. This can be achieved by a high mass transfer rate and a high distribution coefficient of the enzyme between reverse micellar and aqueous phases. The effect of mass transfer coefficient during forward extraction on the recovery of α-amylase was simulated for two values of the distribution coefficient. These model predictions were verified experimentally by changing the distribution coefficient (by adding nonionic surfactant to the reverse micellar phase) and the mass transfer coefficient of the enzyme during the forward extraction (by increasing the **stirrer** speed). The **experimental results** correlated well with the **model predictions**. In this way the performance of the reverse micellar extraction of α-amylase was improved to give an 85% yield of active enzyme in the second aqueous phase and 17-fold concentration of the enzyme. The surfactant losses were also reduced to 2.5% per circulation of the reverse micellar phase. The model also predicted further improvement in the extraction efficiency by **modifying** the extraction techniques, e.g., by reducing the residence time during the extraction in combination with a further increase in the mass transfer rate. The use of **centrifugal separators** or extractors might be valuable in this respect.

Unit 7

New Words

practical application	实际应用
scale up	按比例增加
immiscible phase	不混溶相
diffusion	扩散
distribution coefficient	分配系数
stirrer	搅拌器;搅拌
experimental result	实验结果
model prediction	模型预测法
modify	更改,修改
centrifugal	离心的
separator	离析器,脱脂器

··· Part B ···

Gas Concentration Effects on Secondary Metabolite Production by Plant Cell Cultures

1 Introduction

Plants provide a wide variety of biochemicals useful to humanity. Their uses include medicinal compounds, flavors, fragrances and agricultural chemicals. A number of investigators have studied the use of plant cells in culture, rather than whole plants, as sources of some of the more valuable organic compounds. Before such processes can become a viable manufacturing option, a great deal more must be learned about the optimum conditions for growth and productivity of cells in culture. The use of large quantities of individual cells in a controlled environment is the basis of the well-established fermentation industry. In most fermentation processes, microbial cells (bacteria or fungi) of a particular species are grown in large-scale suspension cultures. Conditions such as temperature, pH and concentrations of dissolved oxygen, carbon source and other nutrients are controlled in order to maximize the production rate of the desired compound. Productivities of commercial fermentation products have been enhanced orders of magnitude above those exhibited in nature. The same principles need to be applied to plant cell cultures.

2 Examples of Secondary Metabolites Produced by Plant Cell Culture

The production of commercial products from plant cell culture processes has been very slow to occur. Although the matter has been discussed in the scientific literature for well over three decades, only a few commercial-scale processes have even been attempted that use plant cell culture to produce secondary metabolites. In Japan three submerged fermentation processes using plant cell cultures were developed by Mitsui Petrochemical Industries to produce berberine, ginseng and shikonin on scales from 4,000 to 20,000 l. Only one of these, production of shikonin by suspension cultures Lithospernum erythrorhizon, can be considered a clear success. It has been operated since the early 1980s. Difficulty in obtaining regulatory approval for use of these products as medicinals has impeded commercialization. In North America commercial production of sanguinarine and vanilla flavor was attempted but failed for regulatory reasons. However, Phyton (Ithaca, NY) utilizes a 75,000 l reactor at a facility near Hamburg, Germany, to develop commercial production of paclitaxel.

In addition to food and fiber, plants are exploited for a large variety of commercial chemicals, including agricultural chemicals, pharmaceuticals, food colors, flavors and fragrances. A few examples of plant-derived pharmaceuticals are digitalis (produced from Digitalis purpurea, prescribed for heart disorders), codeine (Papaver somniferum, sedative), vinblastine and vincristine (Catharanthus roseus, leukemia treatment) and quinine (Cinchona officinalis, malaria). It is believed that plant tissue culture production methods (called "phytoproduction") can be developed to profitably manufacture some of these chemicals. Encouraged by these advantages, Routian and Nickell obtained the first patent for the production of substances by plant tissue culture in 1956. More recent patents cover many aspects of using plant cell culture for secondary metabolite production. Numerous investigators have reported production of useful compounds in hairy root, callus and suspension cultures. The diversity of plant materials adaptable to culturing in hairy root and callus cultures has recently been reviewed. Suspension cultures of Thalictrum minus produced the stomachic and antibacterial berberine. Callus cultures of Stizolobium hassjo produced the antiparkinsonian drug Ldopa. Suspension cultures of Hyoscyamus niger L. produced a derivative of the anticholinergic hyoscyamine.

Some secondary metabolites have been observed in much higher concentrations in cultured cells than in whole plants of the same species. These include ginsengosides from Panax ginseng (27% of cell dry weight in culture, 4.5% in whole plants), anthraquinones from Morinda citrafolia (18% in culture, 2.2% in plants) and shikonin from Lithospernum erythrorhizon (12% in culture, 1.5% in plants).

Additional examples of substances synthesized by cell culture are listed in review articles and books. It is important to note that many of these compounds are secondary metabolites; that is, their production is not related to cell growth and division. Indeed, high rates of production of secondary metabolites usually occur during low rates of growth, often under conditions of significant physiological or biochemical stress on the cells.

3 Comparison with Well-established Microbial Fermentations

Phytoproduction holds several difficulties compared with the well-established microbial fermentations:

1. Plant cells grow much more slowly, with doubling times of the order of 40 h (compared with 0.3 h for some bacteria). Consequently, costs associated with cell generation are much greater.

2. Specific production rate tends to be lower. For example, despite several years of optimization studies, volumetric productivity of shikonin by suspension cultures of L. erythrorhizon was reported as 0.1g product $L^{-1} d^{-1}$. For comparison, the fungus Penicillium crysogenum yields 3.2 g $L^{-1} d^{-1}$ of penicillin.

3. Plant cells may store their products in vacuoles rather than secrete them into the medium.

4. Plant cells are much more sensitive to shear forces than are bacteria or yeast cells, requiring much gentler aeration and agitation.

The medium in contact with cultured plant cells must consist of a large number of components for growth to occur. For the purpose of the present discussion, these components may be categorized as follows:

1. Water
2. Carbon source
3. Concentrated inorganic salts, including nitrogen sources
4. Trace salts (usually considered to be those of less than 0.1 g L^{-1} in the medium)
5. Vitamins
6. Plant hormones and cytokinins
7. Medium conditioning factors (compounds produced in very small quantities by the cells themselves)
8. Dissolved gases

Clearly, the first six component types can be controlled during the initial formulation of the medium. It is these that have been the subject of optimization studies, which form a large part of the recent plant cell culture literature. The seventh category falls outside the capabilities of most investigators in the field. The concentrations of dissolved gases have also been neglected as components, possibly because they cannot be controlled in the same manner as dissolved solids.

4 Importance of Dissoled Gases as Medium Components

In the present study, we attempted to determine the effects of important dissolved gases, oxygen, carbon dioxide and ethylene, on growth of species of three plant genera, Nicotiana, Artemisia and Taxus, in culture. The productivity of respectively valuable metabolic compounds, generic phenolic compounds, artemisinin and paclitaxel, and aspects of culture physiology were also studied. Artemisinin is a promising antimalarial drug and paclitaxel is an effective anticancer drug, so progress towards improved methods of manufacturing would be very desirable. Moreover, N. tabacum, A. annua, T. cuspidata and T. Canadensis can be considered as model systems; the methods (and some of the conclusions) developed from the current studies may be applicable to other phytoproduction systems.

In an efficient commercial phytoproduction process, the cells will be in contact with near optimum concentrations of each dissolved gas. However, to insure that this occurs, the designer must know both the ideal concentrations and (for material balance reasons) the production and usage rates.

5 Specific Consumption of Oxygen

Both the studies of Hulst et al. and Hallsby reported zero-order consumption; that is, in the normal concentration range, cellular O_2 consumption was not dependent on its concentration. LaRue and Gamborg measured O_2 usage of 0.03 mmol gdw^{-1} h^{-1} over the life of a culture of rose cells, but did not investigate the order of reaction.

Taticek et al. tabulated several values of maximum specific O_2 usage rates, which range from $0.2 - 0.6$ mmol gdw^{-1} h^{-1}. Our experiments with A. annua suspension cell cultures indicate a maximum specific usage rate of 0.2 mmol gdw^{-1} h^{-1}. The wide variation in usage rates is suspicious; it may be an artifact of the variety of methods of estimation.

For comparison, several specific usage rates for industrially important bacteria and fungi are given. These range from 3.0 (Aspergillis niger) to 10.8 mmol gdw^{-1} h^{-1} (Escherichia coli).

6 Specific Productivity of Carbon Dioxide

Thomas and Murashige investigated several solid callus cultures of several plant species, but did not test any suspension cultures. Concentrations of CO_2, ethylene and some other gases were measured 24 h after flushing with air. Zobel measured CO_2 and C_2H_4 evolution from soybean callus cultures. Fujiwara et al. measured CO_2 concentrations in closed vessels containing cultured plantlets. None of these reports give enough information to permit the calculation of specific productivity.

7 Specific Productivity of Ethylene

Ethylene (C_2H_4) is produced in essentially every part of every seed plant and affects a number of metabolic functions in very small concentrations. It is therefore considered a plant hormone. Cultured plant cells are also known to produce C_2H_4.

The highest known ethylene release rate is by fading flowers of Vanda orchids, producing approximately $3*10^{-3}$ mmol gdw^{-1} h^{-1}. In some of the sources mentioned in the previous section, both CO_2 and C_2H_4 concentrations were measured, but, again, not enough data is reported to obtain productivity values. LaRue and Gamborg report the amounts of C_2H_4 produced by suspension cultures of several species. The time course data they present shows that C_2H_4 production is very unsteady; however, average productivities can be estimated. For soybean and rose cultures, for example, productivities were $1*10^{-6}$ and $1.4*10^{-5}$ mmol gdw^{-1} h^{-1}, respectively. Most of the other species tested lay between those results. Lieberman et al. measured ethylene productivities of callus and suspension cultures of apple. For callus cultures, productivity peaked at $2.1*10^{-6}$ mmol gdw^{-1} h^{-1}, and averaged about two-thirds that value. For suspension cultures, productivity peaked at $9.4*10^{-6}$ mmol gdw^{-1} h^{-1}, but averaged $2.7*10^{-6}$ mmol gdw^{-1} h^{-1}.

8 Review of Releant Engineering Parameters

In a review article advocating entrapped plant cell cultures, Shuler et al. point out the problem in large-scale suspension cultures of maintaining sufficient oxygen transfer without excessive mechanical shear on the cells. A process designer must not overdesign for oxygen transfer because this would result in increased cell damage. Conventional bioreactors are completely back-mixed stirred tank reactors (STR) that deliver oxygen and remove carbon dioxide by sparging air through the medium. Conventional STR bioreactors exhibit oxygen transfer rates (OTRs) of between 5 – 150 mmol O_2 h^{-1} L^{-1} under normal operating conditions. When the growth-limiting nutrient is oxygen in batch culture, cells grow exponentially until the oxygen uptake (or consumption) rate (OUR) exceeds the oxygen delivery rate (OTR) of the bioreactor. The point at which oxygen becomes the limiting nutrient and the OTR dictates the growth rate can be illustrated at various OTRs using the exponential growth and yield expressions for growth:

$$x_t = x_0 e^{\mu t} \quad (7.1)$$
$$\Delta x = Y_{x/s} \Delta S \quad (7.2)$$

where μ is the specific growth rate (h^{-1}) (which is constant until such time that nutrient becomes limiting, a product becomes toxic, or culture conditions change), t is time (h), x is cell dry weight (g L^{-1}) at the initial condition, 0, and after time t, $Y_{x/s}$ is the yield coefficient, and S is substrate concentration (g L^{-1}).

9 Taxus cuspidata and T. canadensis Production of Paclitaxel: Dependence on Dissolved Gas Concentrations

Paclitaxel is a plant-derived drug used in the treatment of breast, ovarian and lung cancers; clinical trials are underway for treatment of other cancers. The extraction and purification of paclitaxel was initially from Pacific yew trees T. brevifolia and shrubs and trees of other Taxus species: T. baccata, T. cuspidata, T. sumatrana, T. chinensis, T. yunnanensis and T. hicksii. Because the evergreen Taxus brevifolia grows slowly (roughly a foot of height and a half inch of trunk diameter per decade), other techniques were considered to produce the compound without destroying T. brevifolia trees. Bristol-Meyer Squibb is currently manufacturing paclitaxel using a semi-synthesis from 10- deacetylbaccatin III, which is isolated from needles of the Himalayan yew, T. wallinchina.

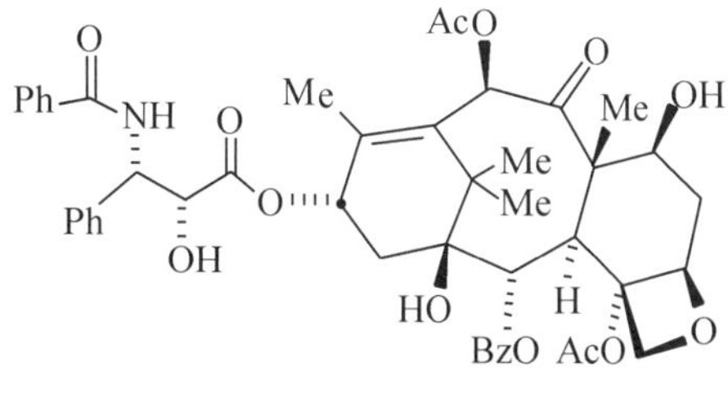

The cell culture process was licensed in May 1995 by Bristol-Meyer Squibb, which in 1998

Fig. 7.3 Structure of Paclitaxel (taxol)

designated $25 million for development of an FDA-approved commercial process. Many academic and industrial research groups around the world are pursuing plant cell culture routes of production.

A significant contribution was made to the science of plant cell culture production of paclitaxel. Methyl jasmonate solutions that had been pipetted into cell suspension cultures caused transient increases in paclitaxel production. This was the first report of modeling that indicated ethylene and methyl jasmonate may participate in "cross-talk" signal transduction in plants. Other papers have subsequently appeared which demonstrate enhancement of paclitaxel productivity by the application of methyl jasmonate to several Taxus species.

The interdependence of methyl jasmonate with chitin- and chitosan-derived elicitors was studied using plant cell suspension cultures of Taxus canadensis. Induction of the biosynthesis of paclitaxel and other taxanes was enhanced when methyl jasmonate and elicitors were added 8 days after culture transfer compared to treatments in which only methyl jasmonate or only elicitor was added. The optimal elicitor concentration using N-acetylchitohexaose was 0.450 μmol L^{-1}, but only in the presence of methyl jasmonate. Little, if any, induction of taxane formation occurred with the oligosaccharide alone. The optimal methyl jasmonate concentration was 200 μmol L^{-1} using colloidal chitin or oligosaccharides of chitin and chitosan as elicitors.

Simple carbohydrates and lipids are proving important as signal induction mediators for regulation of plant growth and development. Fungal cell wall derived oligosaccharides are one group of the former. Methyl jasmonate (MJ) is a lipid-derived elicitor. Both classes of elicitors activate signal transduction pathways and regulate expression of genes for production of phytoalexins and other secondary metabolites. Paclitaxel and other taxanes, such as baccatin III, 10-deacetyltaxol and 10-deacetylbaccatin, are elicited by these materials in Taxus canadensis suspension cultures.

Fatty acid signaling in plants has been reviewed by Farmer. Jasmonic acid arises in plants from α-linolenic acid via the octadecanoic pathway.

Unit 8

··· Part A ···

Established Bioprocesses for Producing Antibodies

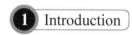

1 Introduction

Following the important development of **hybridoma** technology for producing **monoclonal** antibodies by Kohler and Milstein in 1975, immediate breakthroughs in the treatment of human diseases were expected. The early monoclonal antibodies, originally labeled as "**magic bullets**", faced **clinical** disappointments owing to their **murine** origin which induced the human anti-mouse antibody (HAMA) **immune** response. However, monoclonal antibodies currently represent the second largest single **category** of biopharmaceutical substances under investigation as **therapeutic** drugs. This can be attributed to the recent genetic engineering methods for constructing **chimeric**, **humanized** and human monoclonal antibodies that circumvent the HAMA response. Consequently, the full potential of monoclonal antibodies, with their unique binding **specificity** and potential to be produced in unlimited quantities, is rapidly becoming recognized in biomedical research, **diagnosis** and therapy.

Over the past decade, monoclonal antibodies have been responsible for several of the important advances in pharmacotherapy; agents such as **Synagis**, **Herceptin** and **Remicade** have transformed the treatment of infectious diseases, cancer and **autoimmune** diseases, respectively. However, biopharmaceuticals are among the most expensive of all drugs. For example, the **annual** cost per patient for antibodies like **Rituxan** and **Enbrel** is $10,000 – $15,000. Given the time, cost and risk associated with biopharmaceutical drug development, the ultimate success of new antibody candidates will partly depend on pharmaco economic issues. Increasing pressure from healthcare providers, as well as the much disputed **anticipated** capacity shortage, has triggered a drive to reduce manufacturing costs at the commercial scale by an order of magnitude from $1,000s per gram to $100s per gram. These pressures have encouraged the search for alternative production technologies, such as the use of transgenics, as well as the use of detailed molecular engineering to alter the effectiveness of the antibody. For antibodies to reach their full commercial and medical potential, all improvement efforts need to demonstrate that they can bring down the cost of antibodies. This unit analyses the manufacturing processes used for marketed antibodies for therapeutic and **diagnostic** uses and assesses likely routes for future antibody production.

New Words

hybridoma	杂种瘤;杂种细胞(细胞融合后形成的)
monoclonal	单克隆的;单细胞繁殖的
magic bullets	【药物】魔弹(能够杀灭病毒、癌细胞等又不伤宿主的药物),无副作用的药剂;灵丹妙药;神奇疗法
clinical	临床的;诊所的
murine	鼠科的
immune	免疫的;免于……的,免除的
category	种类,分类
therapeutic	治疗的;治疗学的;有益于健康的
chimeric	嵌合体
humanized	单克隆抗体
specificity	特异性;特征;专一性
diagnosis	诊断
Synagis	帕利珠单抗,为呼吸道合胞病毒融合蛋白(F蛋白)的人单克隆抗体
Herceptin	赫塞汀(抗体药物),重组DNA衍生的人源化单克隆抗体
Remicade	英利昔单抗,是缓解病情的抗风湿药,抑制甲型肿瘤

Unit 8

	坏死因子的药物
autoimmune	自身免疫的;自体免疫的
annual	年度的;每年的
Rituxan	美罗华,是单克隆抗体疗法,已经证实了在 NHL 患者中有效,也被 FDA 批准用于此适应症
Enbrel	依那西普,是肿瘤坏死因子(TNFα)拮抗剂,属 DMAR 药物,是抗风湿病的生物制剂
anticipate	预期,期望;占先,抢先;提前使用
diagnostic	诊断的;特征的

2 Antibody Engineering Efforts

Antibodies are populations of protein molecules, known as **immunoglobulins**, synthesized by an animal in response to a **foreign macromolecule**, an **antigen**. The development of hybridoma technology in mice by Kohler and Milstein allowed, for the first time, the production of monoclonal antibodies recognize specific antigens of choice and led to their widespread application in research and development. However, the use of such murine antibodies as human therapeutics has been limited by their **immunogenicity** in humans (e.g.). Genetic **manipulation** of murine monoclonal antibodies began in the 1980s to reduce their immunogenicity. These techniques were used to generate mouse-human **chimeric antibodies** and then **humanized antibodies** with 90% - 95% human content. They have been shown to be significantly less immunogenic than murine antibodies, and with a longer half-life in the body. More recently, transgenic mice have been genetically engineered to generate fully human antibodies. A number of fully human antibodies are now in clinical development with the aim of eliminating the problems with immunogenicity.

Various antibody fragments can be derived that are proving to be of practical use in **therapy** and diagnosis. Some of these antibody fragments were originally derived from whole antibodies by enzyme **proteolysis**. However, more recent developments in recombinant DNA technology mean that these fragments can also now be produced using expression **hosts** such as Escherichia coli. All of the small fragments (Fv, scFv, Fab, F(ab')2) are useful because they are still capable of binding to antigens and in some therapies this is sufficient. Their smaller size can, in certain situations, improve their diffusion or penetration properties.

New Words

immunoglobulins	免疫球蛋白;免疫球蛋白类
foreign	外来的;异质的,异体的
macromolecule	高分子,大分子
antigen	抗原
immunogenicity	免疫原性;致免疫性;致免疫力
manipulation	操纵;操作;处理
chimeric antibodies	嵌合抗体
humanized antibodies	单克隆抗体
therapy	治疗,疗法
proteolysis	蛋白质水解;蛋白水解作用
host	宿主

3 Investigating Marketed Monoclonal Antibodies

This section provides a summary of the use, nature and manufacture of the approved antibodies investigated. The companies, **indications** and approval dates associated with the antibodies investigated are listed in Tab. 8.1. This information highlighted the fact that antibodies have been approved for the following therapeutic indications: **transplant** rejection; cancer; **cardiovascular** disease; **respiratory** disease; and autoimmune disease. This indicates that antibodies have become successful in clinical trials for a range of indications. Currently, approximately 200 antibodies are in clinical trials. Examining the antibodies in development suggests that in future the major therapeutic applications of antibodies will be for cancer and autoimmune diseases, such as **arthritis**. Most of the diagnostic antibodies approved were for imaging of **cancerous tumour**; additional uses included imaging of cardiovascular disease and infections.

At present all marketed monoclonal antibodies are expressed in **mammalian** cells. This can be attributed to the fact that other cells, such as E. coli, lack the cellular machinery required to secrete antibodies and accomplish necessary **post-translational modifications**, such as **glycosylation**. These modifications are believed to be necessary for various antibody effector

Unit 8

functions and may also influence antibody **half-life** in the body; therefore, the regulatory agencies are concerned about a reproducible glycosylation pattern (e.g.). Analysis of the marketed antibodies revealed that three cell types have been adopted. Murine monoclonal antibodies were produced in hybridoma cells (derived from mouse myeloma cells), whereas genetically engineered antibodies (chimeric or humanized) were produced in either Chinese hamster ovary (CHO) cells or mouse myeloma cells (NS0 or SP2/0 **myeloma cell line**).

Tab. 8.1 Therapeutic monoclonal antibodies and fragments

Product name	Manufacturer	Indication and approval date
Orthoclone OKT3 Muromonab-CD3	Ortho Biotech	Transplant rejection Treatment of acute kidney transplant rejection Treatment of acute heart and liver transplant rejection
Zenapax Daclizumab	Hoffmann-La Roche	Prevention of acute kidney Daclizumab transplant rejection
Simulect Basiliximab	Novartis Pharmaceuticals AG	Prevention of acute kidney transplant rejection Cancer
Rituxan/MabThera Rituximab recombinant	IDEC Pharmaceuticals; Genentech	Treatment of non-Hodgkin's lymphoma
Herceptin Trastuzumab	Genentech	Treatment of metastatic breast cancer
ReoPro Abciximab	Centocor	Prevention of blood clots with high risk coronary angioplasty Prevention of blood clots in refractory unstable angina when coronary intervention is planned
Synagis Palivizumab	MedImmune, Inc. (USA); Boehringer Ingelheim Pharma KG (EU)	Respiratory disease Prevention of respiratory syncytial virus (RSV) infection in paediatrics
Remicade Infliximab	Centocor	Treatment of Crohn's disease Treatment of rheumatoid arthritis

Orthoclone OKT3 was the first monoclonal antibody approved as a biopharmaceutical (for treatment of organ transplant rejection). Following OKT3, subsequent therapeutic antibodies were all genetically engineered and cultured mostly in murine myeloma cells. ReoPro was the first chimeric therapeutic antibody and half of the therapeutic antibodies approved by 1998 were chimeric, produced primarily in mouse myeloma cells rather than CHO cells (three out of four). Rituxan/MabThera was the only chimeric antibody expressed in CHO cells. However, some investigations indicated that both cell types have been widely studied in industry and lend themselves to large-scale processes since they grow in suspension culture.

New Words

indications	适应症
transplant	移植;移植器官;被移植物
cardiovascular	心血管的
respiratory	呼吸的
arthritis	关节炎
cancerous	癌的;生癌的;像癌的
tumour	瘤;肿瘤;肿块
mammalian	哺乳类动物的
post-translational modification	翻译后修饰
glycosylation	糖基化
half-life	半排出期;半留存期
myeloma cell line	骨髓瘤细胞系

4 Generic Antibody Production Processes

From the analysis of the operations commonly used in monoclonal antibody manufacture, potential generic processes can be constructed using either an **affinity**-based process or an ion-exchange-based process with specific **viral clearance** intermediate filtration steps.

Generic processes have the advantage that they greatly reduce the development time and streamline regulatory aspects of processing. Several companies claim to be adopting a generic purification strategy such as Amgen (Thousand Oaks, CA) and Cambridge Antibody Technology

Unit 8

(CAT; Cambridge, UK). Amgen highlights the need to keep robustness in mind when designing a generic purification process. Robustness studies are key to demonstrating that the process performs **adequately** within its control limits and consistently provides material of defined purity, quality and yield. However, CAT cautioned that since all antibodies are slightly different the generic processes just represent a starting point that requires modification for effective scale-up. Industrial-scale antibody production strategies must balance the needs of robustness, purity, yield and economics.

New Words

affinity	[物化][免疫]亲和力
viral	滤过性毒菌引起的;滤过性毒菌的
clearance	清除
adequately	充分地;足够地;适当地

5 Dose and Market Potential as Drivers of Quantities Needed

Reviewing the trends in doses and market potential of therapeutic antibodies is key to assessing what levels of performance new approaches will need to meet to be competitive.

For the therapeutic antibodies, the **cumulative** dose per patient has been derived in this review from clinical studies and product labels. The cumulative price per patient per course of treatment was also found and these data are summarized in Tab. 8.2. This indicated that antibody doses are increasing to gram quantities rather than milligrams, with a concomitant increase in the price of the treatment. All but one of the antibodies require **repeated** dosing. The quantity needed is also a reflection of the indications that antibodies are capable of treating; these now include **chronic** conditions (e.g. Herceptin for **breast** cancer) as well as indications which potentially affect tens to hundreds of thousands of patients (e.g. Remicade for **rheumatoid** arthritis). Tab. 8.2 also contains estimated demands of each drug based on the number of patients to be treated. Attempts were made to derive figures that reflected the potential **market share** captured by each drug. This was achieved by determining the number of patients based on sales reported for each drug. The average annual demand for marketed antibodies was estimated to be 46 kg and is expected to increase as each drug manages to penetrate its market further. It has been claimed that the demand for increasingly large quantities of antibodies is beginning to **outstrip supply of cGMP production capacity**, which could result in antibodies in the development pipeline not reaching the market. However, at present there is considerable controversy regarding the existence of a **shortfall** in worldwide capacity.

Tab. 8.2 Typical cumulative dose, price and market potential of therapeutic antibodies

Product name	Typical cumulative dose/patient(mg)	Typical price/patient/treatment ($)	Potential annual US demand (kg)
Orthoclone OKT3	50 – 70	3,000 – 4,200	0.3 – 0.5
ReoPro	30	1,690	9
Rituxan/MabThera	2,800 – 3,000	14,000 – 15,000	192 – 205
Zenapax	350	6,120	4
Simulect	40	3,160	1
Synagis	600 – 900	7,150 – 8,580	32 – 48
Remicade	350 – 1,050	2,530 – 7,590	20 – 57
Herceptin	5,740	33,350	73

The estimates largely depend on the average **titre** assumed for the currently available cell culture facilities. The disputed **capacity crunch**, along with increasing pressures to drive down the cost of antibody therapy, has led to several developments to address these issues.

New Words

cumulative	累积的
repeated	再三的,反复的
chronic	慢性的;长期的;习惯性的
breast	乳房,胸部
rheumatoid	患风湿症的;风湿症的
market share	市场占有率
outstrip	超过;胜过
shortfall	差额;缺少

titre　　　　　　　　　　　　　　［分化］滴定度；最小滴定度
capacity crunch　　　　　　　　 产能不足

 Future Trends in Antibody Manufacture

Having analyzed processes for approved antibodies, attention is now turned to some of the technologies being used for antibodies in development in order to assess future trends in antibody manufacture.

Regarding cell culture bioreactors, the **Wave Bioreactor** offers an alternative disposable culture system up to 500 L. Singh indicated that batches ranging from 100 mL to 580 L have been run for monoclonal antibodies with productivity and maximal cell densities comparable to those of stirred tank bioreactors. This reactor offers an alternative to the more expensive stainless-steel bioreactors, and has the advantages that come with the use of disposable components, such as not requiring cleaning and **sterilization**. However, the limitations of its scale may **pose** problems during **late phase** clinical trials and launch if scale-up results in a switch in bioreactor technology, as well as the accompanying **bioequivalence** studies. The Wave Bioreactor has also been developed to operate in the **perfusion** mode.

Although mammalian cell culture has emerged as the chosen method of production for marketed antibodies, alternative systems are being developed for antibodies with projected annual marketed demands of several hundreds of kilograms, if not tons. In particular, the use of transgenic plants and animals as culture systems is attracting attention in process development **circles**, for applications requiring very large amounts. **Candidates** for the source of antibodies include the milk of transgenic mammals, the eggs of transgenic chicken, the seeds/leaves/tubers of transgenic plants and the cocoons of silkworms. Increasing interest in these transgenic organisms can be attributed to claims of a competitive cost of goods, a lower capital investment, the flexible capacity and the ability to assemble more complex antibodies, when compared to mammalian cell culture.

Regarding novel approaches in downstream processing, expanded bed **chromatography** has been gaining interest as it can combine clarification, concentration and initial purification in one step. Such processes allow culture broths to be applied directly onto the column without **clogging** it. Industrial, as well as academic, evaluation of expanded bed chromatography has been executed. Affinity chromatography is often the first chromatography step because of its high resolution. However, natural affinity **ligands** have several drawbacks, which include their high cost, their biological origin, ligand **leakage** and poor stability to sanitizing agents. These problems, combined with the fact that fairly large volumes of harvest fluid are loaded onto this column step, have resulted in the emergence of competing alternatives.

New Words

Wave Bioreactor	摇袋式细胞培养生物反应器
sterilization	消毒,灭菌
pose	造成,形成
late phase	延迟相
bioequivalence	生物等效性;生物等量
perfusion	灌注;充满
circle	循环,周期
candidate	候选人,候补者
chromatography	色层分析;色谱分析法
clog	阻塞;障碍
ligand	配体
leakage	泄漏;渗漏物;漏出量

Unit 8

··· Part B ···

The Progress in Biochemical Engineering

① Investigations of the Production of Cephalosporin C by *Acremonium chrysogenum*

Cephalosporins belong together with penicillins to the family of β-lactam antibiotics. In 1945, Giuseppe Brotzu, working at an institute in Caligiari on the island of Sardinia, isolated from seawater near to a sewage outlet a fungus (*Cephalosporium acremonium brotzu*) which produced a mixture of antibiotics. This mixture was effective against both Gram-positive and Gram-negative bacteria, especially against *Staphylococcus aureus* and *Salmonella typhi*. Cephalosporin C (CPC) was isolated and purified by Abraham et al. and its structure was elucidated by Newton and Abraham. *Cephalosporium acremonium* was recently renamed as *Acremonium chrysogenum*. Cephalosporin C is also produced by *Streptomyces clavuligerus* besides clavulanic acid and Cephamycin C produced by *Streptomyces* sp. Five distinguishable morphological forms of *A. chrysogenum* are known: filamentous hyphae, swollen hyphal fragments, arthrospores, conidia and germlings.

Several research groups investigated the biosynthesis of CPC, which is shown in Fig. 8.1.

Because of the close relationship between penicillins and cephalosporins, their biosynthesis is usually discussed jointly. The first step of the biosynthesis is ACV synthesis from cysteine, valine and a-amino adipinic acid (α-AAA) by the enzyme ACV-synthetase (ACVS), which is the bottleneck of the cephalosporin C formation. Therefore, several researchers investigated the regulation of this enzyme. This synthesis is suppressed by ammonia and by phosphate, but is induced by methionine. ATP, L-cysteine and α-AAA act as cofactors and have an activating effect.

The fungus can use several nitrogen sources: amino acids, ammonia, nitrates, urea, etc..

In the industry often $(NH_4)_2SO_4$ is used. Only NH_3 can pass the plasma membrane by free diffusion. ATP consumption is required to the transport of NH_4^+-ion into the cells and to drive out

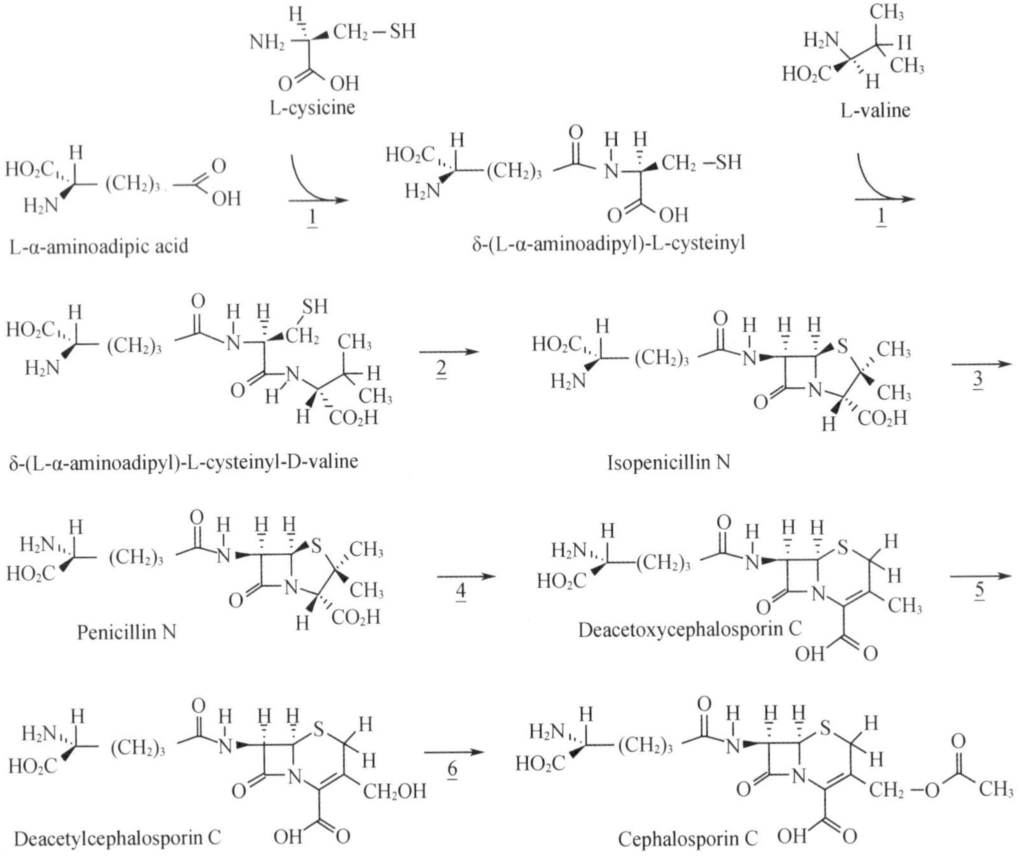

Fig. 8.1 Biosynthesis of cephalosporin C

1. ACV-Synthetase
2. Isopenicillin N-Synthetase
3. Isopenicillin N-Epimerase
4. Deacetoxycephalosporin C-Synthetase
5. Deacetoxycephalosporin C-Hydroxylase
6. Deacetoxycephalosporin C-Acetyltransferase

the protons from the cells. Under the usual cultivation conditions (pH 5.8 – 6.7) the ratio NH_3/NH_4^+ is very low; therefore, diffusion only plays a role at high NH_4OH concentrations.

The application of glucose as a C-source is only recommended during growth, because the easily consumable glucose represses the biosynthesis of cephalosporin C. To avoid this repression, slowly consumable oligo- and poly- saccharides, such as (dry) glucose syrup, starch and dextrin and soy oil and lard oil, respectively, are used in the industry. They are less expensive than

Unit 8

glucose and soy oil has higher energy content and they hinder foam formation.

2 Plasmid Copy Number

What is the meaning of plasmid copy number? The most used definitions are "copies per cell" and "copies per chromosome". Additionally one can find in the literature "fmol plasmid DNA" or "mg plasmid DNA per mg cell dry weight" as indications of quantity. Sometimes comparisons with other plasmid quantities, for example relatively to the copy number of a wild type plasmid serve as specifications for the plasmid copy number.

Plasmids can be classified as low-copy, medium-copy or high-copy plasmids. The ranges of different classes fluctuate in the literature. For low-copy plasmids the range lies between 1 and 10, for medium-copy between 10 to 20 and high-copy plasmids can reach 700 copies and more. Other classifications define 1 – 2 plasmids as low-copy, 6 – 9 as "moderate" and up to 100 as high copy. Different copy numbers occur due to different mechanisms of replication or different genes at the origin of replication. The ori-sequence of the plasmid P15A leads to the low-copy type, whereas plasmids with the ori of pBR322 show a medium copy number. A single point mutation in the gene for RNA I of the replication control of pBR322 leads to the high-copy type. Other mutations increase the copy-number and some models describe the factors influencing copy numbers. For a comparison of such models, see the article by Patnaik.

Normally the copy number is controlled by the regulation of replication. To reach the "normal copy number" after cell division, the replication has to be controlled depending on the actual copy number. That means that the frequency of replication per plasmid corresponds with the total copy number in the cell.

3 Bioprocesses for the Manufacture of Ingredients for Foods and Cosmetics

We all eat and drink processed foods and beverages several thousand times a year, very many of which contain ingredients produced using biocatalysts. Therefore food ingredients, especially for processed foods and beverages, have a big social value as well as creating big business. Ingredient values depend on their "in-product" functionality when formulated into complete foods and beverages. This depends a lot on how an ingredient interacts with the other ingredients, which can affect not only their activities, but also oxidative, microbiological and physical stability and other factors. Therefore assessment of flavour, mouthfeel, visual appearance, microbial and oxidative stability, emulsifying, foaming and phase stabilizing activities, and nutritional properties is essential. This is done by means of manufacturing trials, and shelf-life and taste panel tests etc..

Real and sustained commercial success is only achieved once an ingredient has established a strong advantageous market position over its competitors, both direct and indirect competitors. This advantage can be in one or more of price, functional performance, quality, consumer perception etc.. Therefore in food, beverage and most other industries it is increasingly accepted that most innovations arise directly from "market pull" incentives rather than from "technical push" forces, and that the need for biocatalysts and bioprocesses is actually a derived-demand arising from customer's needs. This means that the market value of the products made by biocatalysts, the costs of production; and the patentability, safety, and consumer acceptability of the product must all be acceptable for the new product to succeed. The importance of "market pull" forces also emphasizes the importance of discovering new functional ingredients and of applications studies to maximize their use in end-products.

The starting point for any process is ingredient that has been traditionally consumed, either because they naturally occur in fruit, cereals or vegetables etc., or are produced in traditional processed foods and beverages. Originally active ingredients such as flavour, colour or antioxidant were only produced in a dilute form, as just one part of the whole food or beverage, such as cheese, butter, yoghurt, beer or wine, depending on the use of Saccharomyces, Lactobacillus strains etc.; and in the East products such as miso, soy sauce and natto using strains such as *Aspergillus oryzae* and *sojai* and *Bacillus natto*. These key ingredients greatly enhance consumer satisfaction and hence the value of these ingredients. Food and cosmetics ingredients include a wide range of different functional materials produced as pure active chemicals (Tab. 8.3). From these traditional products improvements have been made by introducing enzyme steps, particularly for the hydrolysis of proteins, oils and lipids, polysaccharides such as pectins etc. in a wide range of products including brewing, baking, starch processing, dairy and fruit processing etc.. This scientifically-based approach began over 100 years ago, as the first enzyme patent dates from 1894, invented by Takamine for the use of diastatic amylases.

Tab. 8.3　Ingredients of food and cosmetic

Foods	Cosmetics
Emulsifiers	Oils
Stabilizers and thickeners	Emollients

Tab. 8.3 (Continued)

Foods	Cosmetics
Fibre (soluble and insoluble)	Surfactants
Sweeteners	Conditioning polymers
Flavours	UV absorbers
Colours	Skin tanning agen
Preservatives (antioxidants and antimicrobials)	Skin lighteners
Vitamins	Fragrances
Spice extracts	Exfoliating agents
Oils and lipids	Antiwrinkle agen
Protein hydrolysates	Antioxidants
Other micronutrients	Antimicrobials

4 Hematopoietic Stem Cells

Bone marrow (BM) is the principal site for blood cell development in humans. The production of mature blood cells is a continual process that is the result of proliferation and differentiation of stem cells, oligopotent progenitor cells and mature cells (Fig. 8.2). A single stem cell has been proposed to be capable of more than 50 generations (doublings) and has the capacity to generate 10^{15} cells to support up to 60 years of life. The identification, cloning and production of recombinant hematopoietic growth factors/cytokines together with the identification and purification of hematopoietic stem and progenitor cells has enriched our understanding of hematopoiesis. As a result of these fundamental discoveries, many investigators are performing ex vivo manipulation of hematopoietic stem cells (HSC) for potential therapeutic applications, from myeloablative situation to hematological disorders, and from gene therapy to regeneration of specialized tissues.

Dexter et al. described the first in vitro murine BM stem cell culture system. Later, Moore and Sheridan and Gartner and Kaplan adopted the Dexter culture system for human BM culture. During the past two decades several instances of outstanding progress have taken place on ex vivo culture of HSC. It seems likely that, after human skin and cartilage, BM will be the next tissue that will be reconstituted to the point of clinical efficacy. Despite astounding advances in culture of HSC and a fair degree of success in clinical studies, ex vivo expanded hematopoietic cells still remain as a potential source of stem cells for transplantation. The main reason for this is a curious lack of consensus among the investigators that ex vivo expanded HSC could be used in clinical applications. Few investigating groups believed that HSC expand in vitro at the cost of its engraftment potential, so there is no reason why the expanded cells should be used in clinical applications. However, many investigators, by preclinical and clinical studies, demonstrated that ex vivo expanded HSC are well tolerated by the patients; it has marrow repopulating ability and can be differentiated into each type of blood cells.

Several reviews in the recent past have covered various aspects of the ex vivo expansion of HSC and its clinical implications, but none of these reports addressed the basic question— why ex vivo expanded HSC has not yet reached clinics for regular therapeutic usage. In this review we will discuss the technological breakthrough for culture of HSC, biological alterations, and the clinical applications of the expanded cells.

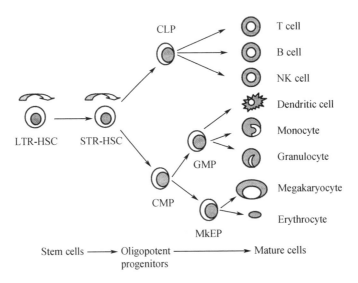

Fig. 8.2 Hematopoietic system hierarchy. Dividing pluripotent stem cells (LTR and STR) may undergo self-renewal to form daughter cells without loss of potential, and also experienced a concomitant differentiation to form oligopotent daughter cells (CLP, CMP, and CMP lead to GMP and MkEP) with more restricted potential. Continuous proliferation and differentiation results in the production of many mature cells. Macrophages are obtained from monocytes and similarly platelets are derived from megakaryocytes (not shown in the figure)

5 Transgenic Birds for the Production of Recombinant Proteins

With the advance in recombinant DNA technology, many therapeutic proteins were successfully produced by bacterial fermentation and animal cell culture using recombinant cells. Transgenic animals such as goat and sheep were engineered to produce large quantities of recombinant proteins in their milk, and the idea is known as transgenic bioreactor. The main advantage of the system is high productivity. It was reported that 35 mg/mL of alpha1-antitrypsin was produced in the milk of the transgenic sheep. In addition, transgenic bioreactor offers a renewable source of highly modified proteins that cannot be produced by microorganisms. However, long breeding time-lines and vast area to maintain flocks are required in such mammals. To overcome the disadvantages, bird species such as chicken and quail have long held promise as alternative transgenic bioreactors. In the case of sheep, it takes 5 and 8 months for gestation and sexual maturation, respectively, while chickens hatch 20 days after oviposition and need 4-6 months for sexual maturation. Furthermore, a typical chicken egg white contains 3-5 g of protein, more than half of which is ovalbumin. Therefore, if ovalbumin in egg white can be partially replaced with a target protein by expressing the transgene in the oviduct of hens, high-yield and low-cost production can be expected. For the realization of this idea, there have been several obstacles including the inefficiency of gene transfer and germ-line transmission, the low hatchability of manipulated embryos, and the absence of an effective transgene expression system. The difficulty in the establishment of transgenic procedures for birds is also due to the complex process of egg formation in hens (Fig. 8.3).

Gene transfer into the avian genome was first achieved in the freshly laid fertilized egg (stage X; blastoderm). Blastodermal cells are pluripotent and contain the precursor cells giving rise to the germ-line. Although blastodermal stage embryos are available in large quantity, the embryo already develops to 60,000 cells. Thus, it is necessary to introduce a transgene into the large number of embryonic cells. For this purpose, bird-derived retroviral vectors have been frequently used for gene transfer. In this case, since the chickens manipulated are mosaic for the transgene, hemizygotes are obtained after mating (Fig. 8.4). However, the germ-line transmission efficiency of transgene was less than 10% possibly due to low viral titer.

In birds, the new technologies for embryo manipulation including embryonic stem (ES) cell and cloning by nuclear transfer have not been established. Besides retroviral gene transfer to blastoderm, microinjection of transgene into single cell stage embryos has been carried out to

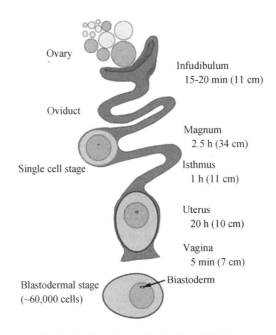

Fig. 8.3 Egg formation in hens birds

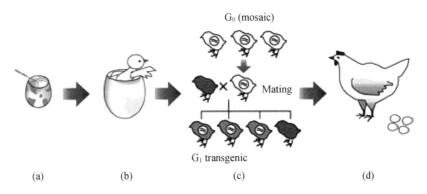

Fig. 8.4 Procedure for generation of transgenic
(a) Injection of gene; (b) Incubation/Embryo culture; (c) Transgenic progeny; (d) Breeding/Production

generate transgenic chickens. However, many problems to be solved exist. For instance, pronuclei cannot be readily visualized in the chicken single cell zygote because of underlying yolk and the opaqueness of the vitelline membrane. In addition, to obtain the freshly fertilized single cell stage embryo in magnum, the donor hen is usually sacrificed. The manipulated embryos must be cultured to hatch by ex vivo culture system using surrogate eggshells. Furthermore, the frequency of transgene insertion into the genome and germ-line transmission efficiency was also very low.

Thus, it is necessary to develop robust and effective techniques to generate transgenic birds. Here we describe our approaches to establish a transgenic bird bioreactor for the production of recombinant pharmaceutical proteins.

6 Development of Culture Techniques of Keratinocytes for Skin Graft Production

In the last decade, the advances in tissue engineering have offered the promising strategies for reconstructing and repairing defected tissues in vivo. In particular, the clinical trials of wound healing have yielded successful outcomes, and several companies are now established as suppliers of tissue engineered products such as skin, cartilage, bone and so on. From a viewpoint of biochemical engineering, however, the culture systems for these tissues are not so sophisticated nor so programmed as compared with the submerged culture systems developed for microorganisms.

In manufacturing the tissue engineered products, the raw materials and final outputs are cells obtained from patients (or donors) and cultures themselves, respectively. The raw materials have heterogeneity and individuality depending on the state of patients and location of excised tissues, and the products vary in size required for patient individuals. These features demand the development of a novel strategy in the manufacturing to obtain the maximum harvests with high quality from the limited supplies of raw materials. For the production of cultured tissues, moreover, the operations including isolation of cells, primary and secondary cultures, and the preparation of cultured tissues are manually performed with tedious and repetitious tasks. The diversities of both raw materials and products as well as the complexity in culture operations oblige us to have a tailor-made process that belongs to the batch operation with less reproducibility.

Fig. 8.5 conceptually compares the manufacturing processes for common and fine chemicals with that for cultured tissues. The manufacturing process for common chemicals is in principle based on large-scale production accompanied with chemical reaction (reaction process) and refining steps through unit operation (purification process). Furthermore, to ensure the purity of products, an additional step of stringent selection for raw materials (sorting process) may be inserted into the manufacturing process especially for fine chemical production. On the other hand, the manufacturing process for autologous cultured tissues has to be constructed on the basis of a concept distinct from the conventional processes described above. The impurities in final products of cultured tissues are difficult to be stripped after reconstruction of tissue. Therefore, the production culture should be achieved with assurance of quality under aseptically controlled conditions. In the case of production for allogenic cultured tissues, to some extent, material selection for sorting might be introducible. For the manufacturing process for the cultured tissues, the engineering problems have not been tackled so far although the studies on kinetic analysis, growth-model development, culture operation, and bioreactor design should be required to perform the reliable and robust process for certification of high-product quality.

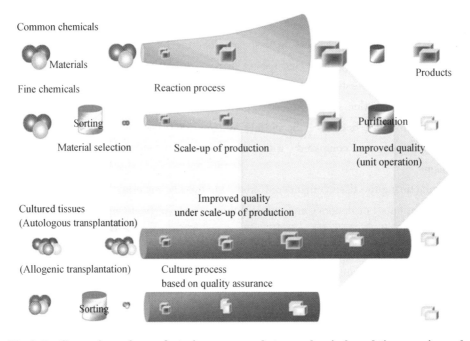

Fig. 8.5 Comparison of manufacturing processes between chemicals and tissue engineered

Cultivation of human keratinocytes is one of the most important steps in the production of epithelial sheet. As shown in Fig. 8.6, the culture operations generally comprise of isolation of keratinocytes from skin biopsy, primary and maintenance cultures, and the sheet formation culture. These cultures accompany with the cellular events of adhesion on culture surface, acclimation after inoculation, growth by cell division, contact inhibition among cells and differentiation accompanied with stratification. The complicated nature owing to these operations and cellular phenomena led to the fluctuation in culture profiles every culture runs.

Unit 8

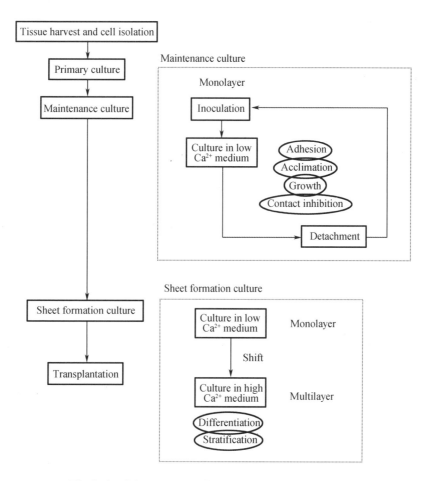

Fig. 8.6 Culture process for epithelial sheet production

References

[1] Suzanne S. Farid. Established Bioprocesses for Producing Antibodies as a Basis for Future Planning[M]. Springer Verlag Berlin Heidelberg ,2006.

[2] Fulton S. Transgenics: the decision matrix. Presented at IBC's 4th international conference: Production and Economics of biopharmaceuticals[C]. San Diego, 2001.

[3] Viktor Nedovic'. Applications of Immobilisation Biotechnology[M]. Springer, Netherlands, 2005.

[4] Yu B, Zhang F, Zheng Y ,et al. Alcohol fermentation from the mash of dried sweet potato with its dregs using immobilised yeast[J]. Process Biochem,1996,31:1 – 6.

[5] Hairston D W. Ethanol:On easy street, for now[J]. Chem. Eng. ,2002,8:36 – 38.

[6] Krishnan M S ,Taylor F, Davison B H, et al. Economic analysis of fuel ethanol production from corn starch using fluidized bed bioreactor[J]. Biores. Technol,2000,75:99 – 105.

[7] Robert A. Copeland. Enzymes: A practical Introduction to Structure, Mechanism, and Data Analysis[M]. 2nd ed. Wiley – VCH, Inc, 2000.

[8] Tsao G T, Brainard A P, Bungay H R, et al. Recent Progress in Bioconversion of Lignocellulosics[M]. Springer, 1999.

[9] Fiechter A. History of Modern Biotechnology I[M]. Springer – Verlag Gmbh, 2000.

[10] Scheper. New Trends and Developments in Biochemical Engineering[M]. Springer, 2004.

[11] Horton R A, Moran L A, Scrimgeour G, et al. Principles of Biochemistry[M]. Prentice Hall, 2002.

[12] Bhatia P K, Danielsson B, Gemeiner P, et al. Thermal Biosensors, Bioactivity, Bioaffinitty[M]. Springer Berlin Heidelberg, 1999.

[13] Scheper T. New Products and New Areas of Bioprocess Engineering[M]. Springer Berlin Heidelberg, 2000.

[14] Dutta N N, Hammar F, Haralampidis K, et al. History and Trends in Bioprocessing and Biotransformation[M]. Springer – Verlag Gmbh, 2002.

References

[15] Zhong J J, Byun S Y, Cho G H, et al. Plant Cells[M]. Springer-Berlin Heidelberg, 2001.

[16] Ulf Stahl, Ute E B. Donalies. Elke Nevoigt, Food Biotechnology[M]. Springer, Berlin Heidelberg, 2008.

[17] Holger Zorn, Peter Czermak. Biotechnology of Foodand Feed Additives[M]. Springer, Berlin Heidelberg, 2014.

[18] Pratyoosh Shukla, Brett I. Pletschke. Advances in Enzyme Biotechnology[M]. Springer, India, 2013.

[19] David P. Clark, Nanette J. Pazdernik. Biotechnology-Applying the Genetic Revolution[M]. Elsevier, 2009.

[20] Hongzhang Chen. Modern Solid State Fermentation[M]. Dordrecht, 2013.

[21] David L. Nelson, Michael M. Cox. Principles of Biochemistry[M]. W H. Freemanand Company, New York, 2008.